3D PRINTING AND ADDITIVE MANUFACTURING

Principles And Applications

(The 5th edition of Rapid Prototyping: Principles and Applications)

3D PRINTING AND ADDITIVE MANUFACTURING

Principles And Applications

(The 5th edition of Rapid Prototyping: Principles and Applications)

Chee Kai Chua

Kah Fai Leong

Nanyang Technological University, Singapore

 **World Scientific**

NEW JERSEY · LONDON · SINGAPORE · BEIJING · SHANGHAI · HONG KONG · TAIPEI · CHENNAI · TOKYO

Published by

World Scientific Publishing Co. Pte. Ltd.

5 Toh Tuck Link, Singapore 596224

USA office: 27 Warren Street, Suite 401-402, Hackensack, NJ 07601

UK office: 57 Shelton Street, Covent Garden, London WC2H 9HE

British Library Cataloguing-in-Publication Data
A catalogue record for this book is available from the British Library.

3D PRINTING AND ADDITIVE MANUFACTURING — Fifth Edition
Principles and Applications
Fifth Edition of Rapid Prototyping

ISBN 978-981-3146-75-4
ISBN 978-981-3146-76-1 (pbk)

Desk Editor: Amanda Yun

Printed in Singapore

To my wife, Wendy, children, Cherie, Clement and Cavell and son-in-law, Darren, whose forbearance, support and motivation have made it possible for us to finish writing this book.

Chee Kai

To Soi Lin, for her patience and support, and Qian, who brings us cheer and joy.

Kah Fai

FOREWORD

Additive manufacturing (AM) and 3D printing, terms used interchangeably, are developing impressively. Countless corporations, government agencies, researchers, and others are investing in AM technology in ways that have not been seen in the past. Many are trying to understand where it is headed and how they fit in. Some of the biggest companies and brands in the world, such as Adobe, Autodesk, Lockheed Martin, Siemens and Stryker have made commitments to AM, some of which are significant. HP, for example, is introducing an entirely new 3D printing technology that is 10 times faster and functions at half the cost of competitive machines.

At the high end of the cost spectrum, companies are qualifying machines and materials for the direct manufacture of parts that go into final products. Aerospace companies such as Aerosud, Airbus, Bell Helicopter, Boeing and Honeywell Aerospace are qualifying AM processes and materials and certifying new designs for flight. In April 2016, GE Aviation delivered to Airbus its first two LEAP engines that include 3D printed fuel nozzles. The advanced design consolidates 20 parts into one, is 25% lighter and five times more durable.

When consolidating many parts into one or fewer parts, companies eliminate part numbers, inventory, assembly, labour, inspection, maintenance and certification paperwork. Using topology optimisation — a method of letting mathematics decide where to put the material to optimise the strength to weight ratio — it is possible to produce parts that are up to 50% lighter. Also, it is possible to redesign parts with relatively thin skins with internal lattice, mesh or cellular structures instead of using solid material throughout. This can further reduce the amount of material and weight of a part.

The AM industry continues to encounter challenges. Among them are system reliability and process repeatability, especially when using AM for manufacturing. AM system manufacturers are addressing these challenges with real-time process monitoring and control software, but a lot of work remains. The current limitations in build speed and maximum part size are challenges, too. A few manufacturers are developing systems with larger build volumes and methods that increase throughput.

Potential users of AM and others have misconceptions about the technology. For example, significant differences exist between industrial-grade systems and the sub-$5000 3D printers. For example, most AM systems are relatively slow, and economies of scale apply, but not in the same way as conventional methods of manufacturing. Also, AM processes are generally more energy-intensive than conventional processes. AM is not a "push button" technology, and pre- and post-build work is usually required, especially for metal parts. AM is not inherently superior to subtractive or formative methods of manufacturing, although it depends on the types of parts being built and the design optimisation that went into them.

The industries that are among the early adopters of AM for series production include aerospace, medical, dental and some consumer products. AM favours the types of parts that are produced in relatively low volumes, are of high value and complex. As system speed increases, overall costs decline, making AM viable for a broader range of production applications. One example is automotive, which has used AM for two decades for prototyping, but most companies are not currently using it for production applications. This will change in the future.

It is an exciting time for design and manufacturing. Never before have so many new options become available. The AM industry has had 28 years to learn how to apply the technology to modelling and prototyping. For the most part, AM is well understood for these applications. The next frontier is to apply AM to the production of parts for final products. This is where the largest opportunities lie and where most investments will be made in the future.

Terry Wohlers
Wohlers Associates, Inc.

PREFACE

The focus on productivity has been one of the main concerns of industries worldwide since the turn of new millennium. To increase productivity, industries have attempted to apply more computerised automation in manufacturing. Amongst the latest technologies to have significant stride over the past two and a half decades are the *Rapid Prototyping (RP) Technologies*, otherwise also known as *3D Printing* or, according to ASTM standards, now formally known as *Additive Manufacturing (AM)*.

The revolutionary change in factory production techniques and management requires a direct involvement of computer-controlled systems in the entire production process. Every operation in this factory, from product design, to manufacturing, to assembly and product inspection, is monitored and controlled by computers. CAD-CAM or Computer-Aided Design and Manufacturing has emerged since the 1960s to support product design. Up to the mid-1980s, it has never been easy to derive a physical prototype model, despite the existence of CNC (or Computer Numerical Controlled) machine tools. Early RP Technologies provided that bridge from product conceptualisation to product realisation in a reasonably quick manner, without the fuss of NC programming, jigs and fixtures.

In recent years, there has been a growing trend in the development and use of AM. The Economist has forecasted that AM could be the third industrial revolution due to its prospects of thriving into a new type of manufacturing industry. Manufacturing is an established and necessary sector in the spectrum of any world's economy. It makes significant contributions to a country's gross domestic production and national security through the constant supply of essential products and services.

From a local perspective of how the world adopts and adapts to AM, there is no reason for Singapore to stay stagnant. Indeed, there is a little internal pressure for Singapore to focus on AM development due to her relatively smaller market, but the external push is very real. First world countries, such as the United States, the United Kingdom and Australia have already established national centres or institutes on AM at various scales, setting an example for other advanced countries. In Asia, China is closely following this technological trend and has attempted to develop its own AM brands. However, the East is still far behind the West in general. Singapore, as a place where East meets West and a place where AM can be rooted back to the late 1980s, must take the leadership in developing new AM-based manufacturing technologies and training future AM engineer leaders, for Singapore as well as for the Asia Pacific region.

With this exciting promise, the industry and academia have internationally established research centres for AM, with the objectives of working in this leading edge and disruptive technology, as well as of educating and training more engineers in the field of AM. On that same premise, the Singapore Centre for 3D Printing (SC3DP) (formally known as Nanyang Technological University (NTU) AM Centre) was founded in July 2013 to be the regional leader in Asia Pacific in AM technologies. Currently, there are seven AM processes categorised under the ASTM standard and each process has its unique advantages and challenges. All of these are covered in this book.

The SC3DP focuses on the powder bed fusion process, which is one of the most accepted and (in the view of some, the) most important AM processes. However, much of this process is still under research today. The SC3DP initiative aims to realise its full potential. SC3DP will tap on the existing local AM experience and expertise to set up a large scale AM facility for research and education to generate new knowledge, create novel technologies and train manpower, especially in the area of Selective Laser Sintering, Selective Laser Melting and Electron Beam Melting. In addition, a conducive AM environment akin to the SC3DP favours the development and building of AM capabilities and strengths.

Beyond that, these can be directly translated and extended into the industry for a variety of applications. By taking advantage of Singapore's position and manufacturing reputation in the aerospace and marine industries, coupled with anticipating infrastructures and an accessible pool of talents, SC3DP offers an ideal setting in Asia to attract and support global businesses by focusing on three core sectors: Research, Education and Application.

In order to continue the focus on education in AM, this textbook, already in its fifth edition, is therefore needed as the basis for the development of a curriculum in AM. Like its previous four editions, the purpose of this updated book is to provide an introduction to the fundamental principles and application areas in AM. The book traces the development of AM in the arena of Advanced Manufacturing Technologies and explains the principles underlying each of the common AM techniques. Also covered are the detailed descriptions of the AM processes and their specifications. In this fifth edition, new AM techniques are introduced and existing ones updated, bringing the total number of AM techniques described to more than 20. To reflect the new terminology and for consistency, we have replaced "Rapid Prototyping" which was used from the first to third edition to "Additive Manufacturing". "3D Printing" will sometimes be used as a common term to refer to "Additive Manufacturing". The book would not be complete without emphasising the importance of AM applications in manufacturing and other industries. Thus an entire chapter is devoted to application areas. As AM has expanded its scope of applications, one whole chapter that focuses on the biomedical arena has been added. New applications include food printing, wearable fashion, garment printing, weapon printing and 3D printing in building and construction.

One key inclusion in this book is the use of multimedia to enhance the reader's understanding of the technique. In the accompanying Companion Media Pack, video and animation are used to demonstrate the working principles of major AM techniques such as Stereolithography, Polyjet, Selective Deposition Lamination, Fused Deposition Modelling, Selective Laser Melting and Selective Laser

Sintering. In addition, the book focuses on some of the very important issues facing AM today, and these include, but are not limited to:

1. The problems with the *de facto* STL format and emerging new standards like AMF and 3MF to replace STL.
2. The range of applications for tooling and manufacturing, including biomedical engineering.
3. The benchmarking methodology in selecting an appropriate AM technique.

The material in this book has been used, revised and updated several times for professional courses conducted for both academia and industry audiences since 1991. Certain materials were borne out of research conducted in the School of Mechanical and Aerospace Engineering at the Nanyang Technological University, Singapore.

To be used more effectively for graduate or final year (senior year) undergraduate students in Mechanical, Aerospace, Production or Manufacturing Engineering, problems have been included in this textbook. For university professors and other tertiary-level lecturers, the subject of AM can be combined easily with other topics such as: CAD, CAM, Machine Tool Technologies and Industrial Design.

Chua C. K.
Professor

Leong K. F.
Associate Professor

School of Mechanical and Aerospace Engineering
Nanyang Technological University
50 Nanyang Avenue
Singapore 639798

ACKNOWLEDGEMENTS

First, we would like to thank God for granting us His strength throughout the writing of this book. Secondly, we are especially grateful to our respective spouses, Wendy and Soi Lin, our respective children, Cherie, Clement, Cavell Chua and Leong Qianyu and son-in-law Darren (Cherie's husband) for their patience, support and encouragement throughout the year it took to complete this edition.

We wish to thank the valuable support from the administration of Nanyang Technological University (NTU), especially the School of Mechanical and Aerospace Engineering (MAE) in particular the NTU Additive Manufacturing Centre (NAMC). In addition, we would like to thank our current and former students and research fellows, Dr. An Jia, Dr. Khalid Rafi Haludeen Kutty, Dr. Alex Liu, Dr. Zhang Danqing, Chan Lick Khor, Chong Fook Tien, Chow Lai May, Esther Chua Hui Shan, Derrick Ee, Foo Hui Ping, Witty Goh Wen Ti, Ho Ser Hui, Ketut Sulistyawati, Kwok Yew Heng, Angeline Lau Mei Ling, Jordan Liang Dingyuan, Liew Weng Sun, Lim Choon Eng, Liu Wei, Phung Wen Jia, Mahendra Suryadi, Mohamed Syahid Hassan, Anfee Tan Chor Kwang, Tan Leong Peng, Toh Choon Han, Wee Kuei Koon, Yap Kimm Ho, Yim Siew Yen, Yun Bao Ling, Zhang Jindong, and our colleague, Mr. Lee Kiam Gam for their valuable contributions that made the multimedia pack possible.

We would also like to express sincere appreciation to our special assistants Chu Shih Zoe, Lu Zhen, Vu Trong Thien, Ng Chyi Huey, Wang Yawei, Kum Yi Xuan, Chen Zi Yang, Huang Sheng, Timothy Yap Shi-Jie and Xu Ming for their selfless help and immense effort in the coordination and timely publication of this book.

Much of the research work which have been published in various journals and now incorporated into some chapters in the book can be attributed to our former students and colleagues, Dr. May Win Naing, Dr. Yeong Wai Yee, Dr. Florencia Edith Wiria, Dr. Novella Sudarmadji, Dr. Tan Jia Yong, Dr. Jolene Liu, Dr. Liu Dan, Dr, Zhang Danqing, Tong Mei, Micheal Ko, Ang Ker Chin, Tan Kwang Hui, Liew Chin Liong, Gui Wee Siong, Simon Cheong, William Ng, Chong Lee Lee, Lim Bee Hwa, Chua Ghim Siong, Ko Kian Seng, Chow Kin Yean, Verani, Evelyn Liu, Wong Yeow Kong, Althea Chua, Terry Chong, Tan Yew Kwang, Chiang Wei Meng, Ang Ker Ser, Lin Sin Chew, Ng Chee Chin and Melvin Ng Boon Keong, and colleagues Professor Robert Gay, Mr. Lee Han Boon, Dr. Jacob Gan, Dr. Du Zhaohui, Dr. M. Chandrasekeran, Dr. Cheah Chi Mun, Dr. Daniel Lim Chu Sing and Dr. Georg Thimm.

We would also like to extend our special appreciation to Mr. Terry Wohlers for his foreword, and Professor Richard Buswell of Loughborough University, Professor Michael C McAlpine of Princeton University, Associate Professor Hod Lipson of Cornell University, Professor Ming Leu of the University of Missouri-Rolla, Dr. Amba D. Bhatt of the National Institute of Standards and Technology, Gaithersburg, MD, Ruston, and Professor Fritz Prinz of Stanford University, USA for their assistance.

The acknowledgements would not be complete without the contributions of the following companies for supplying and helping us with the information about their products they develop, manufacture or represent:

1. 3D Systems Inc., USA
2. 3D-Micromac AG, Germany
3. Arcam AB, Sweden
4. Aeromet Corp., USA
5. Alpha Products and Systems Pte Ltd., Singapore
6. Blizzident, USA
7. CAM-LEM Inc., USA
8. Carl Zeiss Pte Ltd., Singapore

9. Champion Machine Tools Pte Ltd., Singapore
10. CMET Inc., Japan
11. Concept Laser GmbH, Germany
12. Creatz3d Pte Ltd., Singapore
13. Cubic Technologies Inc., USA
14. Cubital Ltd., Israel
15. Cybron Technology (S) Pte Ltd., Singapore
16. Defense Distributed, USA
17. EnvisionTec., Germany
18. Ennex Corporation, USA
19. EOS GmbH, Germany
20. Festo, Germany
21. Fraunhofer-Institute for Applied Materials Research, Germany
22. Fraunhofer-Institute for Manufacturing Engineering and Automation, Germany
23. Innomation Systems and Technologies Pte Ltd., Singapore
24. Kira Corporation, Japan
25. MCOR Technologies, Ireland
26. MIT, USA
27. Materialise, Belgium
28. Optomec Inc., USA
29. RegenHU Ltd., Switzerland
30. SLM Solutions, Germany
31. Solid Concepts, USA
32. Solidimension Ltd., Israel
33. Solidscape Inc., USA
34. Soligen Inc., USA
35. Stratasys Inc., USA
36. The ExOne Company, USA
37. ThreeASFOUR, USA
38. voxeljet AG, Germany
39. Zugo Technology Pte Ltd, Singapore

We would also like to express our special gratitude to the following individuals who have agreed to showcase their 3D printing work and creations in our book: Cody R. Wilson, Olaf Diegel, Amit Zoran, Neri Oxman and Jeffrey Lipton.

Last but not least, we also wish to express our thanks and apologies to the many others not mentioned above for their suggestions, corrections and contributions to the success of the previous editions of the book. We would appreciate your comments and suggestions on this fifth edition book.

Chua C. K.
Professor

Leong K. F.
Associate Professor

ABOUT THE AUTHORS

CHUA CK is the Executive Director of the Singapore Centre for 3D Printing (SC3DP) and a full professor of the School of Mechanical and Aerospace Engineering at Nanyang Technological University (NTU), Singapore. Over the last 25 years, Professor Chua has established a strong research group at NTU, pioneering and leading in computer-aided tissue engineering scaffold fabrication using various additive manufacturing (AM) techniques. He is internationally recognised for his significant contributions in the biomaterial analysis and rapid prototyping process modelling and control for tissue engineering. His work has since extended further into AM of metals and ceramics for defence applications.

Professor Chua has published extensively with over 300 international journals and conferences, attracting over 6000 citations, and has a Hirsch index of 37 in the Web of Science. His book, *3D Printing and Additive Manufacturing: Principles and Applications*, now in its fifth edition, is widely used in American, European and Asian universities and is acknowledged by international academics as one of the best textbooks in the field. He is the World No. 1 Author for the area of AM and 3D Printing (or rapid prototyping as it was previously known) in the Web of Science and is the most "Highly Cited Scientist" in the world for that topic. He is the Co-Editor-in-Chief of the international journal, *Virtual & Physical Prototyping* and serves as an editorial board member of three other international journals. In 2015, he started a new journal, the International Journal of Bioprinting and is the current Chief Editor. As a dedicated educator who is passionate in training the next generation, Professor Chua is widely consulted on AM (since 1990) and has conducted more than 60 professional development courses for the industry and academia in Singapore and the world. In 2013, he was awarded the "Academic Career Award" for his contributions to AM (or 3D Printing) at the 6th International Conference on Advanced Research in Virtual and Rapid Prototyping (VRAP 2013), 1–5 October 2013, at Leiria, Portugal. Professor Chua can be contacted by e-mail at mckchua@ntu.edu.sg.

LEONG KF is an associate professor at the School of Mechanical and Aerospace Engineering, NTU, Singapore. He has taught for more than 30 years in NTU in a variety of courses, including Creative Thinking and Design, Product Design and Development, Industrial Design, AM, Collaborative Design and Engineering, amongst others. He graduated from the National University of Singapore and Stanford University for his undergraduate and graduate degrees respectively. His principal areas of research interests are in AM and its applications, particularly in the biomedical areas, including tissue engineering and drug delivery devices. He has also research interests in sports technology, product design, product development science and design education. He has chaired several committees in the Singapore Institute of Standards and the Industrial Research and Productivity Standards Board, receiving Merit and Distinguished Awards for his services in 1994 and 1997, respectively. He has delivered keynote papers at several conferences on the application of AM in biomedical science and tissue engineering. He has authored over 180 publications, including books, book chapters, international journals and conferences and has two patents to his name. He has won several academic prizes and his publications have been cited more than 4700 times. His H-index is 29 based on the Science Citation Index. He was the Head of the Systems and Engineering Management Division in the School of Mechanical and Aerospace Engineering and the Founding Director of the Design Research Centre at the University. He is the Programme Committee Deputy Chair of the SC3DP and also the Co-Director of the SIMTech-NTU Joint Laboratory on 3D AM. Concurrently, he is the director of the Institute for Sports Research, a research institute in NTU. Leong Kah Fai can be contacted by e-mail at mkfleong@ntu.edu.sg.

LIST OF ABBREVIATIONS

2D	=	Two-Dimensional
3D	=	Three-Dimensional
3DP	=	Three-Dimensional Printing
ABS	=	Acrylonitrile Butadiene Styrene
ACS	=	Advanced Composite Structures
ACES	=	Accurate Clear Epoxy Solid
AIM	=	ACES Injection Moulding
AM	=	Additive Manufacturing
AMF	=	Additive Manufacturing File
AOM	=	Acoustic Optical Modulator
ASCII	=	American Standard Code for Information Interchange
ASTM	=	American Society for Testing and Materials
BBR	=	Federal Office for Building and Regional Planning
BDM	=	Beam Delivery System
CAD	=	Computer-Aided Design
CAE	=	Computer-Aided Engineering
CAGR	=	Compound Annual Growth Rate
CAM	=	Computer-Aided Manufacturing; Ceramics Additive Manufacturing
CAM-LEM	=	Computer-Aided Manufacturing of Laminated Engineering Materials
CBC	=	Chemically Bonded Ceramics
CDSA	=	Combat Direction Systems Activity
CIM	=	Computer-Integrated Manufacturing
CIMP-3D	=	Centre for Innovative Metal Processing through Direct Digital Deposition
CLC	=	Closed-Loop Control
CLI	=	Common Layer Interface

CLIP	=	Continuous Liquid Interface Production
CJP	=	ColorJet Printing
CMET	=	Computer Modelling and Engineering Technology
CMM	=	Coordinate Measuring Machine
CNC	=	Computer Numerical Control
CSG	=	Constructive Solid Geometry
CT	=	Computerised Tomography
CTTC	=	Centre for Technology Transfers in Ceramics
DARPA	=	Defense Advanced Research Projects Agency
DLP	=	Digital Light Processing
DSO	=	Defence Science Organisation
DMD	=	Direct Metal Deposition; Digital Micromirror Device
DMLS	=	Direct Metal Laser Sintering
DoD	=	Drop on Demand
DPM	=	Digital Part Materialization
DSP	=	Digital Signal Processor
DSPC	=	Direct Shell Production Casting
DTI	=	Danish Technological Institute
EB	=	Electron Beam
EBAM	=	Electron Beam Additive Manufacturing
EBM	=	Electron Beam Melting
ECC	=	Engineering Cementitious Composites
EDM	=	Electric Discharge Machining
ERM	=	Enhanced Resolution Module
ELI	=	Extra Low Interstitial
EU	=	European Union
FCP	=	Freeze Cast Process; Fast Ceramic Production
FDM	=	Fused Deposition Modelling
FEA	=	Finite Element Analysis
FEM	=	Finite Element Method
FRP	=	Fibre-Reinforced Polymer
GE	=	General Electrics
GIS	=	Geographic Information System
GPS	=	Global Positioning System
HPGL	=	Hewlett–Packard Graphics Language
HQ	=	High Quality

HR	=	High Resolution
HS	=	High Speed
HVAC	=	Heating. Ventilation and Air Conditioning
IGES	=	Initial Graphics Exchange Specification
IR	=	Integrated Resources
IRISS	=	Interlayer Realtime Imaging and Sensing System
ISO	=	International Standards Organisation
LAN	=	Local Area Network
LCD	=	Liquid Crystal Display
LDW	=	Laser Deposition Welding
LEAF	=	Layer Exchange ASCII Format
LED	=	Light Emitting Diode
LENS	=	Laser Engineered Net Shaping
LMD	=	Laser Metal Deposition
LMT	=	Laser Manufacturing Technologies
LOM	=	Laminated Object Manufacturing
LS	=	Laser Sintering
LWD	=	Linear Weld Density
MAE	=	Mechanical and Aerospace Engineering
M^3D	=	Maskless Mesoscale Material Deposition
M-RPM	=	Multi-Functional RPM
MEM	=	Melted Extrusion Modelling
MEMS	=	Micro-Electro-Mechanical Systems
MIG	=	Metal Inert Gas
MJF	=	Multi-Jet Fusion
MJM	=	Multi-Jet Modelling System
MJP	=	Multi-Jet Printing
MJS	=	Multiphase Jet Solidification
MRI	=	Magnetic Resonance Imaging
NA	=	Numerical Aperture
NASA	=	National Aeronautical and Space Administration
NAVAIR	=	Naval Air Systems Command
NAVSEA	=	Naval Sea Systems Command
NSWCDD	=	Naval Surface Warfare Center Dahlgren Division
NC	=	Numerical Control
ND	=	Neutral Density

NIR	= Near Infrared	
NTU	= Nanyang Technological University	
OPC	= Open Packaging Conventions	
OPO	= Optical Parametric Oscillator	
PC	= Personal Computer; Polycarbonate	
PCB	= Printed Circuit Board	
PDA	= Personal Digital Assistant	
PLT	= Paper Lamination Technology	
PPSF	= Polyphenylsulfone	
PSDO	= Partner Standards Developing Organisation	
PSI	= Phillips Service Industries, Inc.	
PSL	= Plastic Sheet Lamination	
PTI	= Plastic Sheet Lamination	
RDM	= Thermoplastic Polyetherimide	
RE	= Reverse Engineering	
ReM	= Reference Model	
RFP	= Rapid Freeze Prototyping	
RM&T	= Rapid Manufacturing and Tooling	
RP	= Rapid Prototyping	
RPI	= Rapid Prototyping Interface	
RPM	= Rapid Prototyping and Manufacturing	
RPS	= Rapid Prototyping Systems	
RPT	= Rapid Prototyping Technologies	
RSP	= Rapid Solidification Process	
RM&T	= Rapid Manufacturing and Tooling	
SAHP	= Selective Adhesive and Hot Press	
SBIR	= Small Business Innovative Research	
SCS	= Solid Creation System	
SD	= Standard Deviation	
SDL	= Selective Deposition Lamination	
SDM	= Shaped Deposition Manufacturing	
SFF	= Solid Freeform Fabrication	
SFM	= Solid Freeform Manufacturing	
SGC	= Solid Ground Curing	
SHCC	= Strain-Hardened Cement-based Composites	
SHR	= Single Head Replacement	

SLA	=	Stereolithography Apparatus
SLC	=	Stereolithography Contour
SLM	=	Selective Laser Melting
SLS	=	Selective Laser Sintering
SMS	=	Selective Mask Sintering
SPCTS	=	Science of Ceramic Processes and Surface Treatment
SRAM	=	Static Random Access Memory
SSM	=	Slicing Solid Manufacturing
SOUP	=	Solid Object Ultraviolet-Laser Plotting
STEP	=	Standard for the Exchange of Product
STL	=	Stereolithography File
3MF	=	3D Manufacturing Format
TPM	=	Tripropylene Glycol Monomethyl Ether
TPP	=	Two-Photon Polymerisation
TTL	=	Toyota Technical Centre
UAM	=	Ultrasonic Additive Manufacturing
UC	=	Ultrasonic Consolidation
UV	=	Ultraviolet
UAM	=	Ultrasonic Additive Manufacturing
UAS	=	Unmanned Aerial System
UAV	=	Unmanned Aerial Vehicle
VM	=	Virtual Model

CONTENTS

CHAPTER 1

INTRODUCTION

The competition in the world market for manufactured services and products has intensified tremendously in recent years. It has become important, if not vital, for new products to reach the market as early as possible, before the competitors.[1,2] To bring products to the market swiftly, many of the processes involved in the design, testing, manufacture, marketing and distribution of the product have been squeezed, both in terms of time and material resources. The efficient use of such valuable resources calls for more efficient tools and effective approaches in dealing with them, and many of these tools and approaches have evolved. They are almost entirely technology-driven, inevitably involving the computer, a result of the rapid development and advancement in such technologies over the last few decades.

Additive manufacturing (AM), formerly known as rapid prototyping (RP), is one such technological development. AM is defined by ASTM International, formerly known as the American Society for Testing and Materials, as "a process of joining materials to make objects from 3D model data, usually layer upon layer, as opposed to subtractive manufacturing methodologies".[3] Also commonly known as 3D printing in public literature, this emerging technology is revolutionising the manufacturing industry with its ability to turn digital data into physical parts. Its distinct ability to manufacture complex shapes and structures has already made it invaluable for the production of prototypes such as engine manifolds for the automotive industry and tools such as investment casting moulds in the jewellery and aeronautical industries. As the technology becomes more developed, AM is moving on to the direct production of components and parts. The potential of AM has resulted in an increase in public interest in recent years, especially with

1

the media coverage of novel and sometimes controversial applications of AM, such as the production of printable guns and food printing.[4] This book follows the previous two editions: *3D Printing and Additive Manufacturing: Principles and Applications* (The 4th edition of *Rapid Prototyping: Principles and Applications*) by Chua and Leong (2014, with companion media pack)[5] as well as *Rapid Prototyping: Principles and Applications*, 3rd edition by Chua et al. (2010),[6] and will provide a clear, comprehensive and updated coverage of this exciting new technology.

1.1 Prototype Fundamentals

According to the Oxford Advanced Learner's Dictionary of Current English: a prototype is the first or original example of something that has been or will be copied or developed[7]; it is a model or preliminary version. A general definition in design terms would be an approximation of a product (or system) or its components in some form for a definite purpose in its implementation.

Prototypes are used for the following purposes:

(1) Experimentation and learning while designing
(2) Testing and proofing of ideas and concepts
(3) Communication and interaction among design teams
(4) Synthesis and integration of the entire product concept
(5) Scheduling and markers

Prototypes created with AM technologies can serve most if not all of these roles.

Prototypes can be broadly classified by three aspects: implementation, form and degree of approximation.

1.1.1 *Implementation*

Complete prototypes will model most of the characteristics of the final product. The prototypes are usually full-scale and they are also usually fully functional.

Alternatively, prototypes can be partially functional in nature. These prototypes are either needed to study and investigate special problems associated with a single component or sub-assembly, or they are needed to study and validate a concept that requires close attention.

1.1.2 *Form*

Prototypes can either be virtual or physical.

Virtual prototypes are non-tangible, and they are used when the physical prototype are too large, too expensive or too time-consuming to produce. Virtual prototypes are based purely upon the assumed principles or science at that point in time, and are completely unable to predict any unexpected phenomenon. Nowadays, virtual prototypes can be stressed, tested, analysed and modified as if they were physical prototypes. An example is the visualisation of airflow over an aircraft wing.

Physical prototypes are tangible, and are built for the purpose of testing, experimentation or aesthetic and human factors evaluation. Physical prototypes can be manufactured using AM techniques, or other methods which tend to be less sophisticated, craft-based and largely labour-intensive. An example is the mock-up of a mobile phone.

1.1.3 *Degree of approximation*

Prototypes can be very rough representations of the intended final product. The prototype is therefore used for testing and studying certain problems that can arise from product development.

Alternatively the prototype can be an exact full-scale representation of the product, modelling every aspect of the product. This type of prototype becomes more and more important towards the end-stage of the product development process.

1.2 Historical Development

In 1987, the first commercial AM system, stereolithography apparatus (SLA)-1, was launched by 3D Systems in the United States.[8] It worked on the principle of stereolithography (STL) and for the first time enabled users to generate a physical object from digital data. The invention of AM technology was a "watershed event"[9] because of the tremendous time saved by not machining parts, especially for complicated and difficult-to-produce models. Since then, other new AM technologies have been commercialised, including Fused Deposition Modelling (FDM) (see Section 4.1) and Selective Deposition Lamination (SDL) (see Section 4.2) in 1991, as well as Selective Laser Sintering (SLS) (see Section 5.1) in 1992.

New applications for AM systems have also been developed and expanded as users gained more experience with the machines. Improvements in materials, such as new resins with better mechanical properties when cured, and technologies like more accurate and faster building techniques, have also made AM a feasible way to produce tooling. In 1993, Soligen brought Direct Shell Production Casting to the market and 3D Systems also introduced QuickCast (a method of producing indirect tooling from AM), as the potential savings in time and resources that AM can bring became apparent. Companies such as General Motors were early adopters of AM. They acquired the SLA-250 (the model immediately following SLA-1) in 1991, and used it for rapid tooling and prototyping of parts such as cranking motor nose housings and connector feeder tracks.[10]

New AM technologies continued to emerge and old ones became more refined as Kira Corp introduced a system that works on paper lamination in 1994; 3D printers based on technology similar to inkjet printers began to appear in the market in 1996. In 1998, Optomec sold its first Laser Engineered Net Shaping (LENS) metal powder system (see Section 5.4) as the LENS process was capable of producing fully dense metal parts with no voids within the metal. In 1999, selective laser melting (SLM) system (see Section 5.7) was introduced by Fockele & Schwarze of Germany. Since 2000, there have been many more AM systems entering the market. As more AM systems capable of producing fully dense metal parts emerged, there was growing interest in refining the process of producing metal parts directly for the aerospace and automobile industries. While some companies such as Boeing were already using AM to produce parts such as electrical boxes, brackets and environmental control system ducting,[8] the development of international standards was expected to help different companies coordinate AM research and commercialisation efforts and further increase the use of AM for direct manufacturing. AM has developed from a tool, first for "visualisation", then to RP, before being applied to produce tooling and eventually becoming able to build parts with sufficient properties allowing it to be used for direct manufacturing.

New AM applications and capabilities are emerging all the time, including applications in bioengineering which led to the development of machines such as 3D Discovery and BioFactory from RegenHU Ltd (see Section 3.6). Both machines use the proprietary BioInk which is a semi-synthetic hydrogel that can support cell growth. New machines with new functionalities such as multi-material, multi-colour printing are being released.

The expiration of older patents, such as the thermoplastic extrusion technology patent used by the FDM (expired in 2009), has led to an explosion of development of an array of low-cost personal 3D printers such as machines by MakerBot, RepRap and so on. These low-cost machines can even be self-assembled from spare parts or kits, and allow individuals to access AM technology at a relatively low price. These machines have also spawned online communities where AM design files are shared on open sources. Since then, many other AM patents have also expired, peaking in 2014, and this has led to a greater variety of low-cost printers flooding into the consumer market space. All these hint at a future where AM will become an integral part of everyday life.

1.3 Fundamentals of AM

Regardless of the different techniques used in the AM systems developed, they generally adopt the same basic approach, which can be described as follows:

(1) A model or component is modelled on a Computer-Aided Design and Computer-Aided Manufacturing (CAD/CAM) system. The model, which describes the physical part to be built, must be represented as closed surfaces which unambiguously define an enclosed volume. This means that the data must specify the inside, the outside and the boundary of the model. This requirement will become redundant if the modelling technique used is based on solid modelling. This is by virtue that a valid solid model will automatically have an enclosed volume. This requirement ensures that all horizontal cross sections that are essential to AM are closed curves to create the solid object.

(2) The solid or surface model to be built is next converted into a format called the .STL file format which originated from 3D Systems. The STL file format approximates the surfaces of the model using the simplest of polygons and triangles. Highly curved surfaces must employ many triangles, and this means that STL files for curved parts can be very large. However, some AM systems also accept data in the IGES (Initial Graphics Exchange Specification) format, provided it is of the correct "flavour".

(3) A computer program analyses a STL file that defines the model to be fabricated and "slices" the model into cross sections. The cross sections are systematically recreated through the solidification of either liquids or powders and then combined to form a 3D model. Another possibility is that the cross sections are already thin, solid laminations and these thin laminations are glued together with

adhesives to form a 3D model. Other similar methods may also be employed to build the model.

Fundamentally, the development of AM can be described in four primary areas. The AM Wheel in Fig. 1.1 depicts these four key aspects of AM. They are: input, method, material and applications.

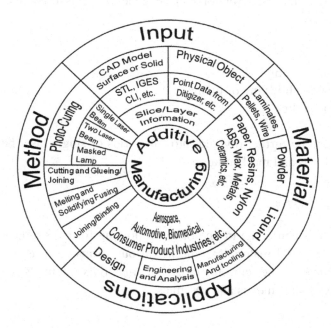

Figure 1.1: The additive manufacturing wheel depicting the four major aspects of AM.

1.3.1 *Input*

Input refers to the electronic information required to describe the object in 3D. There are two possible starting points — a computer model or a physical model or part. The computer model created by a CAD system can be either a surface model or a solid model. On the other hand, 3D data from the physical model is not so straightforward. It requires data acquisition through a method known as reverse engineering. Equipment, such as coordinate measuring machines (CMM) or laser digitisers, are used to capture data points of the physical model, usually in a raster format, and then to "reconstruct" it in a CAD system.

1.3.2 *Method*

While they are currently more than 50 vendors for AM systems, the method employed by each vendor can be generally classified into the following categories:

(1)   Photo-curing
(2)   Cutting and joining
(3)   Melting and solidifying or fusing
(4)   Joining or binding

Photo-curing can be further divided into categories of single laser beam, double laser beams and masked lamp.

1.3.3 *Material*

The raw materials can come in one of the following forms: solid, liquid or powder state. Solid materials come in various forms such as pellets, wire or laminates. The current range of materials includes paper, polymers, wax, resins, metals and ceramics.

1.3.4 *Applications*

Most of the AM parts are finished or touched up before they are used for their intended applications. Applications can be grouped into (1) design, (2) engineering analysis and planning and (3) manufacturing and tooling. A broad range of industries can benefit from AM and these include, but are not limited to, aerospace, automotive, biomedical, consumer, electrical and electronic products.

1.4 Advantages of AM

Today's automated, tool-less, pattern-less AM systems can directly produce functional parts in small production quantities. Parts produced in this way usually have an accuracy and surface finish inferior to those made by machining. However, some advanced systems are able to produce near tooling quality parts that are close to or are in the final shape. The parts produced, with appropriate post-processing, have material qualities and properties close to the final product. More importantly, the time taken to manufacture any part — once the design data are available — is short, and can be a matter of hours.

The benefits of AM systems are immense and can be broadly categorised into direct and indirect benefits.

1.4.1 *Direct benefits*

There are many benefits for the company using AM systems. One would be the ability to experiment with physical models of any complexity in a relatively short time. In the last 40 years, products realised to the market place have become increasingly complex in shape and form.[11] For instance, compare the aesthetically beautiful car body of today with that of the 1970s. On a relative complexity scale of 1–3, as seen in Fig. 1.2, it can be noted that from a base of 1 in 1970, this relative complexity index has increased to about 2 in 1980, approached 3 in the 1990s, and exceeded 3 after 2000. However, the relative project completion times have not correspondingly increased. It increased from an initial base of about 4 weeks' project completion time in 1970 to 16 weeks in 1980. However, with the use of CAD/CAM and computer numerical control (CNC) technologies, project completion time was reduced to 8 weeks. Eventually, AM systems allowed the project manager to further cut the completion time to less than 2 weeks in 2015.

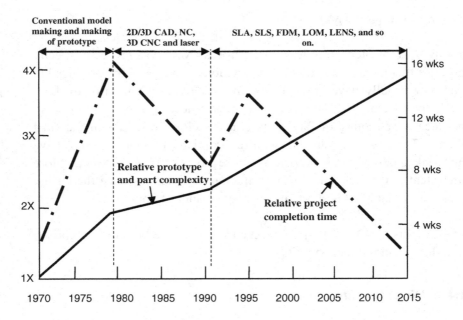

Figure 1.2: Project time and product complexity in 40 years' time frame.

To the individual in the company, the benefits can be different and have varying impact. It depends on the role they play in the company. The full production of any product encompasses a wide spectrum of activities. Kochan and Chua[12] describe the impact of AM technologies on the entire spectrum of product development and process realisation in Fig. 1.3.

The activities required for full production in a conventional model are depicted at the top of Fig. 1.3. The bottom of Fig. 1.3 shows the AM model. Depending on the size of production, savings on time and cost could range from 50% to 90%.

1.4.1.1 *Benefits to product designers*

Product designers can increase part complexity with little effect on lead time and cost. More organic, sculptured and complex shapes for functional or aesthetic features can be accommodated. They can optimise part design to meet customer requirements, with few restrictions imposed

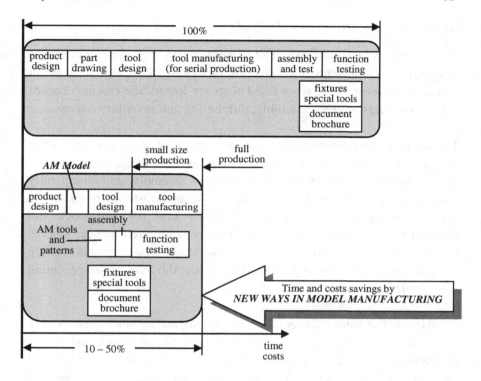

Figure 1.3: Results of the integration of AM technologies.

by manufacturing. In addition, they can reduce parts count by combining features into single-piece parts that were previously made from several because of poor tool accessibility or the need to minimise machining and waste. With fewer parts, the time spent on tolerance analysis, selecting fasteners, detailing screw holes and assembly drawings is greatly reduced.

There will also be fewer constraints on that parts can be designed without regards to draft angles, parting lines or other such constraints. In other words, parts can be now designed with accurate, large and thin walls and can be machined without the use of stock shapes. Product designers can minimise the use of material and optimise strength/weight ratios without regards to machining cost. Finally, they can minimise time-consuming discussions and evaluations of manufacturing possibilities.

1.4.1.2 *Benefits to tooling and manufacturing engineers*

The main savings are in costs. The manufacturing engineer can minimise design, manufacturing and verification of tooling. She can realise profit earlier on new products, since fixed costs are lower. She can also reduce parts count and therefore assembly, purchasing and inventory expenses.

The manufacturer can reduce the labour content of manufacturing, since part-specific setting up and programming are eliminated, machining or casting labour is reduced, and inspection and assembly are consequently minimised as well. Reducing material waste, waste disposal costs, material transportation costs and inventory cost for raw stock and finished parts (producing only as many parts as required reduces storage requirements) can contribute to lower overheads. Fewer inventories are scrapped because of fewer design changes, and the risks of disappointing sales are reduced.

In addition, the manufacturer can simplify purchasing since there is only one form of raw material: a spool of wire, a vat of liquid, and so on. Quotations vary little among suppliers, since fabrication is automated and standardised. The production manager can purchase one general purpose machine rather than many specialised machines, and therefore reduce capital equipment and maintenance expenses and minimise the need for specialised operators and training. A smaller production facility will also make scheduling production easier. Furthermore, one can reduce the inspection reject rate since the number of tight tolerances required where parts must mate can be reduced. One can avoid design misinterpretations (instead, "what you design is what you get"), quickly change design dimensions to deal with tighter tolerances and achieve higher part repeatability, since tool wear is eliminated. Lastly, one can reduce spare parts inventories (produce spares on demand, even for obsolete products).

1.4.2 *Indirect benefits*

Outside the design and production departments, indirect benefits can also be derived. Marketers as well as consumers will benefit from the utilisation of AM technologies.

1.4.2.1 *Benefits to marketing and supply chain*

To the market, it presents new capabilities and opportunities. It can greatly reduce time-to-market, resulting in (1) reduced risk as there is no need to project customer needs and market dynamics several years into the future, (2) products which better satisfy customers' needs, (3) products offering the price/performance of the latest technology, (4) new products being test-marketed economically.

AM can change production capacity according to market demand, possibly in real time, with little or no impact on manufacturing facilities. One can increase the diversity of product offerings and pursue market niches which are currently too small to justify due to tooling cost (including custom and semi-custom production). One can easily expand distribution and quickly enter new markets.

AM can also potentially simplify the supply chain for certain types of businesses, resulting in lower logistic costs. Unlike traditional manufacturing practices, multiple AM machines can be operated by a single person, reducing labour costs. Its ability to produce complex parts at no extra cost may lead to new designs with fewer separate components and hence, reduce costs incurred from product assembly. Moreover, the AM machine may be located close to the assembly line, eliminating shipping cost and time. The lack of dependence on cheap labour can result in the reduction of outsourcing, resulting in manufacturing returning to the countries from which the designs originate. All these will shorten and simplify the supply chain,[13] reducing costs of production and time taken to reach consumers. It also means that businesses may need to rethink their current business models, and that new business models with a focus on innovation and mass customisation may become more prevalent.

1.4.2.2   *Benefits to the consumer*

The consumer can buy products which are customised to suit individual needs and wants. Firstly, since there are fewer constraints on the manufactured parts, there is a much greater diversity of offerings to choose from. Secondly, one can buy (and even contribute to the design of) affordable built-to-order products. Furthermore, the consumer will enjoy lower prices, since the manufacturers' savings will ultimately be passed on.

1.5   Commonly Used Terms and Definitions of AM

There are many terms used by the engineering communities around the world to describe this technology. Perhaps this is due to the versatility and continuous development of the technology. In the public domain, the most commonly used term is *3D printing*. Although this term within the industry describes only the fabrication of objects through the deposition of material using a print head or nozzle, it is now more commonly used in a non-technical context to refer to AM.[3] Previously, the most commonly used term was *RP*. The term was apt as AM could *rapidly* create a physical model. However, this technology is increasingly being used in areas other than prototyping, and so the term is now considered outdated.[8]

Some of the less commonly used terms include *Direct CAD Manufacturing*, *Desktop Manufacturing* and *Instant Manufacturing*. The rationale behind these terms is also based on AM's speed, ease and convenience, though in reality it is hardly direct or instant. *CAD Oriented Manufacturing* is another term and provides an insight into the origin of the model used for manufacturing.

Another group of terms emphasises the unique characteristic of AM — layer-by-layer addition of material. This group includes *Layer Manufacturing, Material Deposit Manufacturing, Material Addition Manufacturing* and *Material Incress Manufacturing*. This is in contrast to more traditional manufacturing methods, such as machining, which describes various material removal methods.

There is yet another group of terms which chooses to focus on the word "freeform" — *Solid Freeform Manufacturing* and *Solid Freeform Fabrication*. *Freeform* stresses the ability of AM to build complex shapes with little constraint on its eventual form. Such constraints in form will apply to traditional machining methods — for instance, it is difficult if not impossible to create a void in a solid object using machining methods.

The term AM has been chosen by the F42 committee of ASTM International to describe this process in the ASTM F2792 standard.[14] This standard has in 2015 been superseded by the ISO/ASTM 52900[3] jointly published by ASTM International and the International Organization for Standardization (ISO). This standard includes definitions for general terms, applications and so on.

1.6 Classification of AM Systems

ISO/ASTM, in its most recent ISO/ASTM 52900 General Principles — Terminology standards manual, has classified all AM processes into seven broad categories.[3] Binder jetting is a process in which a liquid bonding agent is selectively deposited to join powder materials. Directed energy deposition allows focused thermal energy to fuse materials by melting as they are deposited. Material extrusion is a process where material is selectively dispensed through a nozzle or orifice. Material jetting occurs when droplets of build material like photopolymer and wax are selectively deposited. Powder bed fusion is a process where thermal energy selectively fuses regions of a powder bed. In sheet lamination, sheets of material are bonded to form a part. Finally, in vat photopolymerisation, liquid photopolymer in a vat is selectively cured by light-activated polymerisation.

Another way to broadly classify the numerous AM systems in the market is by the initial form or phase of its material. Due to the large number of proprietary systems, this book categorises select AM systems into (1) liquid-based, (2) solid-based and (3) powder-based systems.

1.6.1 *Liquid-based*

The initial form of liquid-based AM systems' building material is the liquid state. Through a process commonly known as curing, the liquid is converted to the solid state. The following AM systems fall into this category:

(1) 3D Systems' SLA
(2) Stratasys' Polyjet
(3) 3D Systems' Multijet Printing (MJP)
(4) EnvisionTEC's Perfactory
(5) CMET's Solid Object Ultraviolet-Laser Printer (SOUP)
(6) EnvisionTEC's Bioplotter
(7) RegenHU's 3D Bioprinting
(8) Rapid Freeze Prototyping

As illustrated in the AM Wheel in Fig. 1.1, there are three possible methods under the *Photo-curing* method. The *single laser beam* method is a widely used method and includes two of the above AM systems: (1) and (5). (2) and (3) both use UV lamps for curing after deposition of photocurable polymers via jetting heads. (4) uses an imaging system called Digital Light Processing (DLP) while both (6) and (7) use the extrusion method in a liquid medium for creating the objects. (8) involves the freezing of water droplets and deposits in a manner much like FDM (see the following) to create the prototype. These are described in more details in Chap. 3.

1.6.2 *Solid-based*

Except for powder, solid-based AM systems are meant to encompass all forms of material in the solid state. In this context, the solid form can include shapes like wires, rolls, laminates and pellets. The following AM systems fall into this definition:

(1) Stratasys' FDM
(2) Mcor Technologies' SDL
(3) Sciaky's Electron Beam AM (EBAM)
(4) Fabrisonic's Ultrasonic AM (UAM)

Referring to the AM Wheel in Fig. 1.1, two methods are possible for solid-based AM systems. AM systems (1), (3) and (4) belong to the *Melting and Solidifying or Fusing* method, while the *Cutting and Joining* method is used for AM systems (2). The various AM systems will be described in more detail in Chap. 4.

1.6.3 *Powder-based*

In a strict sense, powder is by-and-large in the solid state. However, it is intentionally created as a category outside the solid-based AM systems to mean powder in grain-like form. The following AM systems fall into this definition:

(1) 3D Systems' SLS
(2) 3D Systems' ColorJet Printing (CJP)
(3) EOS's EOSINT Systems
(4) Optomec's LENS
(5) Arcam's Electron Beam Melting (EBM)
(6) Concept Laser GmbH's LaserCUSING®
(7) SLM Solutions GmbH's SLM®
(8) 3D Systems' Phenix PXTM
(9) 3D-Micromac AG's MicroSTRUCT
(10) The ExOne Company's ProMetal
(11) Voxeljet AG's VX System

All the above AM systems employ the *Joining/Binding* method. The method of joining/binding differs for the above systems in that some employ a laser while others use a binder/glue to achieve the joining effect. Similarly, the above AM systems will be described in more detail in Chap. 5.

References

1.	Wheelwright, SC and KB Clark (1992). *Revolutionizing Product Development: Quantum Leaps in Speed, Efficiency, and Quality*. New York: The Free Press.

2.	Ulrich, KT and SD Eppinger (2011). *Product Design and Development*, 5th Ed. Boston: McGraw Hill.

3.	ISO/ASTM 52900:2015, Additive manufacturing — General Principles — Terminology. ISO/ASTM International, Switzerland. 2015.

4.	Greenberg, A (May 2013). $25 Gun Created With Cheap 3D Printer Fires Nine Shots (Video). *Forbes*. http://www.forbes.com/sites/andygreenberg/2013/05/20/25-gun-created-with-cheap-3d-printer-fires-nine-shots-video/

5.	Chua, CK and KF Leong (2014). *3D Printing and Additive Manufacturing: Principles and Applications (The 4th edition of Rapid Prototyping: Principles and Applications)*. Singapore: World Scientific Publishing.

6.	Chua, CK, KF Leong and CS Lim (2010). *Rapid Prototyping: Principles and Applications,* 3rd Ed. Singapore: World Scientific Publishing.

7.	Prototype [Def 1] (2016). *Oxford Advanced Learner's Dictionary*. Oxford: Oxford University Press. http://www.oxfordlearnersdictionaries.com/definition/english/prototype?q=prototype

8.	Wohlers, T and T Caffrey (2013). History of additive manufacturing. In *Wohlers Report 2013*, pp. 15–17. Colorado: Wohlers Associates.

9.	Kochan, D (1992). Solid freeform manufacturing: possibilities and restrictions. *Computers in Industry*, 20, 133–140.

10.	Jacobs, PF (1993). *Stereolithography 1993: Epoxy Resins, Improved Accuracy & Investment Casting*. Dearborn, Michigan.

11. Metelnick, J (1991). *How Today's Model/Prototype Shop Helps Designers Use Rapid Prototyping to Full Advantage.* Society of Manufacturing Engineers Technical Paper MS91-475.

12. Kochan, D and CK Chua (1995). State-of-the-art and future trends in advanced rapid prototyping and manufacturing. *International Journal of Information Technology*, 1(2), 173–184.

13. Tuck, C, R Hague, and N Burns (2007). Rapid manufacturing: impact on supply chain methodologies and practice. *International Journal of Services and Operations Management*, 3(1), 1–22.

14. ASTM F2792, Standard Terminology for Additive Manufacturing Technologies. ASTM International, West Conshohocken, PA, USA. 2009.

Problems

1. Describe the historical development of AM and related technologies.

2. Despite the increase in relative complexity of product's shape and form, project time has been kept relatively shorter. Why?

3. What are the fundamentals of AM?

4. What is the *AM Wheel*? Describe its four primary aspects. Is the *Wheel* a static representation of what AM is today? Why?

5. Describe the advantages of AM in terms of its benefits to product designers, tool designers, manufacturing engineers, marketers and consumers.

6. Name three AM Systems that are liquid based.

7. How can a liquid form be converted to a solid form using these liquid-based AM Systems?

8. What form of materials for AM Systems can be classified as solid based? Name three such systems.

9. What is the mechanism behind powder-based AM systems?

10. Many terms have been used to refer to AM. Discuss three of such terms and explain why they have been used in place of AM.

11. Give a few examples of the applications of AM.

12. What are the implications of AM technology for different industries?

CHAPTER 2

ADDITIVE MANUFACTURING PROCESS CHAIN

2.1 Fundamental Automated Fabrication Processes

There are three fundamental fabrication processes[1,2] as shown in Fig. 2.1. They are *Subtractive, Additive and Formative* Fabricators.

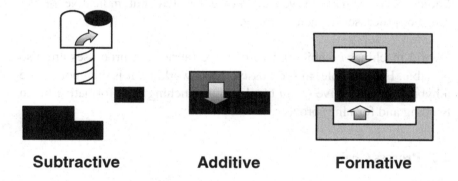

Subtractive   **Additive**   **Formative**

Figure 2.1: Three types of fundamental fabrication processes.

In a subtractive process, one starts with a single block of solid material larger than the final size of the desired object. Portions of the material are removed until the desired shape is reached.

In contrast, an additive process is the exact reverse of subtractive process. The end product is much larger than the initial material. Materials are manipulated so that they are successively combined to form the desired object.

Lastly, formative process is one where mechanical forces or restricting forms are applied on a material so as to form it into the desired shape.

There are many examples of each of these fundamental fabrication processes. Subtractive fabrication processes include most forms of machining processes — computer numerical control (CNC) or otherwise. These include milling, turning, drilling, planning, sawing, grinding, electric discharge machining (EDM), laser cutting, water-jet cutting and the like. Most forms of 3D printing and additive manufacturing (AM) processes such as stereolithography apparatus (SLA) and selective laser sintering (SLS) fall into the additive fabrication processes category. Examples of formative fabrication processes are: bending, forging, electromagnetic forming and plastic injection moulding. These include both bending of sheet materials and moulding of molten or curable liquids. The examples given are not exhaustive but indicative of the range of processes in each category.

Hybrid machines combining two or more fabrication processes are also possible. For example, in progressive press-working, it is common to see a hybrid of subtractive (as in blanking or punching) and formative (as in bending and forming) processes.

2.2 Process Chain

As described in Section 1.3, all additive AM techniques adopt the same basic approach. As such, all AM systems generally have a similar sort of process chain. Such a generalised process chain is shown in Fig. 2.2.[3] There are a total of five steps in the chain and these are: Step 1: 3D modelling and support design; Step 2: data conversion and transmission, Step 3: checking and preparing, Step 4: building and Step 5: post-processing. Depending on the quality of the model and part in steps 3 and 5, respectively, the process may be iterated until a satisfactory model or part is achieved.

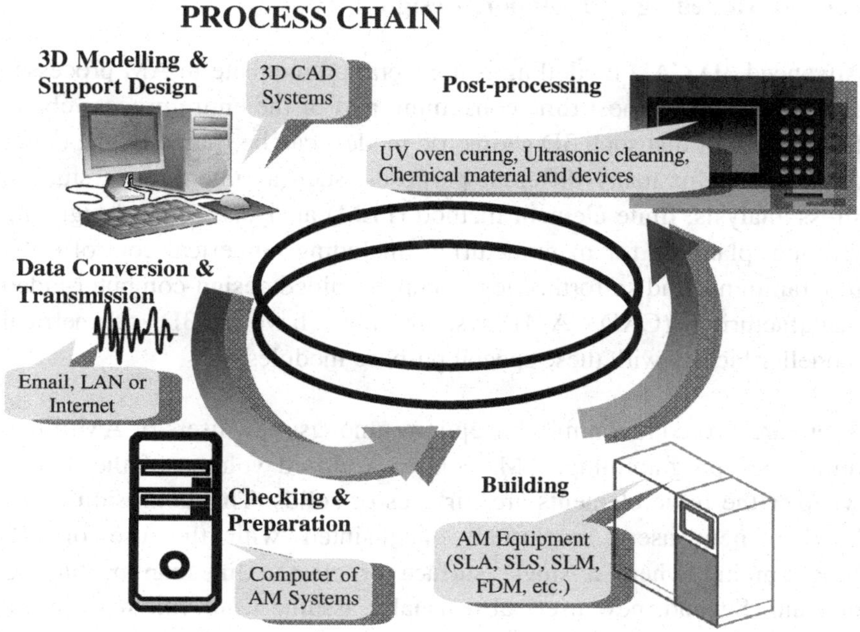

PROCESS CHAIN

Figure 2.2: Process chain of additive manufacturing systems.

However, like other fabrication processes, process planning is important before AM commences. In process planning, the steps of the AM process chain are listed. The first step is 3D geometric modelling and support design. In this instance, the requirement would be a computer and a Computer-Aided Design (CAD) modelling system. The factors and parameters, which influence the performance of each operation are examined and decided upon. For example, if SLA is used to build the part, the orientation of part is an important factor which would, amongst others, influence the quality of the part and the speed of the process. Thus an operation sheet used in this manner requires proper documentation and sound guidelines. Good documentation, such as a process logbook, allows future examination and evaluation and subsequent improvements can be implemented in process planning. The five steps are discussed in the following sections.

2.3   3D Modelling and Support Design

Advanced 3D CAD modelling is a general prerequisite in AM processes and is usually the most time-consuming part of the entire process chain. It is important that such 3D geometric models can be shared by the entire design team for many different purposes, such as interference studies, stress analysis, finite element method (FEM) analysis, detail design and drafting, planning for manufacturing, including numerical control (NC) programming, and so forth. Most computer-aided design-computer-aided manufacturing (CAD-CAM) systems now have a 3D geometrical modeller facility with these special purpose modules.

There are two common misconceptions amongst new users of AM. First, unlike NC programming, AM requires a closed volume of the model, whether the basic elements are surfaces or solids. This confusion arises because new users are usually acquainted with the use of NC programming where a single surface or even a line can be an NC element. Second, new users also usually assume *what you see is what you get*. These two misconceptions often lead to the user under-specifying process parameters to AM systems, resulting in poor performance and non-optimal utilisation of the system. Examples of considerations that have to be taken into account include orientation of parts, supports for the part, difficult-to-build part structures such as thin walls, small slots or holes and overhanging elements. Therefore, AM users have to learn and gain experience by working on the AM system. The problem is usually more complex than one can imagine because there are many different AM machines, which have different requirements and capabilities.

Based on the misconceptions of *what you see is what you get*, overlooking of the importance of support design is not uncommon. AM has indeed enabled the possibility for previously impossible to fabricate structures. However, practically speaking, the desired geometrical freedom is limited by the overhanging structure and many other problems. Fortunately, the limitation can be unbounded by the means of support.

The term "support" in the AM context, fittingly, is loosely defined as a structure or material, which supports a 3D printed part to prevent adverse structural effects arising mainly from, but not limited to, gravity, where such support will normally be removed at the end of an AM process. Its main purposes are to (1) prevent deformation, curl, dross formation and so forth due to overhanging structure (see Fig. 2.3), (2) anchor floating object (see Fig. 2.4) and (3) separate part from the platform to facilitate adhesion and removal of part on and from the platform (see Fig. 2.4). Concurrently, in some cases of metallic parts, support also facilitates heat transfer during the printing process to mitigate the effects of residual stresses arising from thermal cycle.[4]

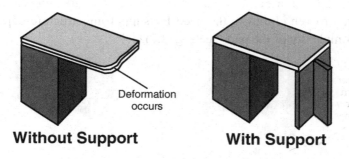

Figure 2.3: Overhanging structure.

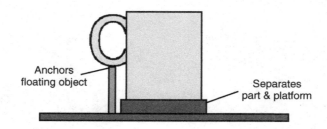

Figure 2.4: Support anchoring floating object and separating part from platform.

The means of generating support structure can be done manually or more often, with the aid of appropriate software. For example, Statysys Inc's FDM has created an intelligent software known as the SupportWorks™, which will automatically design the necessary supports given a part's geometry. It is also important to know that some other technologies do not require, or have a minimal requirement on supports, such as SLS and digital light processing (DLP), whereby taking SLS as an example, the unsintered powder automatically serves as the natural support.

Moreover, the support structure can be reduced depending on the design. Thus, it would be good to mention, albeit briefly, on this aspect. For instance, a heavier overhanging structure of a part will require a stronger support structure, thereby increasing the total material needed (see Fig. 2.5). Meanwhile, usage of angled support allows self-supporting of overhanging structure, therefore reducing the need of having a support (see Fig. 2.6). Furthermore, a good choice of part orientation can minimize or even eliminate the need for supporting structure depending on the complexity of the part (see Fig. 2.7).

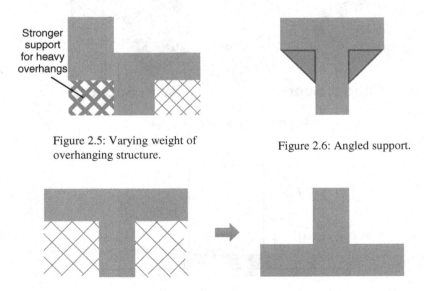

Figure 2.5: Varying weight of overhanging structure.

Figure 2.6: Angled support.

Figure 2.7: Elimination of support structure with different orientation.

The 45° rule is the most common rule of thumb for approximating whether support is needed to print a part. The rule dictates the usage of supporting structure for any overhanging parts having less than 45° angle from the horizontal axis (see Fig. 2.8). However, this rule is not universal, as the requirement of support structure differs with different 3D printing technologies and material choices. Nevertheless, the rule of thumb provides an insight on the shortcoming of AM that needs to be addressed.

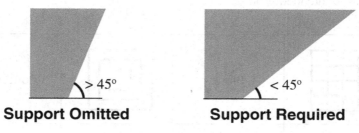

Support Omitted   **Support Required**

Figure 2.8: The 45° rule of thumb.

It is essential to note that different AM technologies deal with support differently. To better understand the various aspects of support requirements, categorising them might be an easier way to describe them. The various aspects of AM supports may be categorised as following:

(1) *Reusable versus. Non-reusable Supporting Material.*
    In most instances, support structures are deemed as a waste at the end of an AM process, as the support structure is not readily reusable for the next AM process. A series of recycling processes are necessary to convert the support structure back to the original form of the raw material again. In contrast, a powder-based polymer AM machine uses its raw material as its support that can be removed and reused at the end of the AM process, thus reducing the need for support structure and waste generation. Another example of reusable supporting material can be observed in the rapid freeze prototyping (RFP) technology, whereby the supporting structure is composed of eutectic sugar solution that can be easily melted and reused.

(2) *The Geometry of Support Structure.*

There are many possibilities for the support structure design geometry. Possible support geometry for metal parts includes, but not limited to block, point, web, contour, line and tube (see Fig. 2.9), in which the block and line support structures are more commonly used.[5] Each type of support geometry has its purposes with the primary goal to minimise waste generation, ease of removal of support structure/material and the impact on the surface finishing of the intended product.

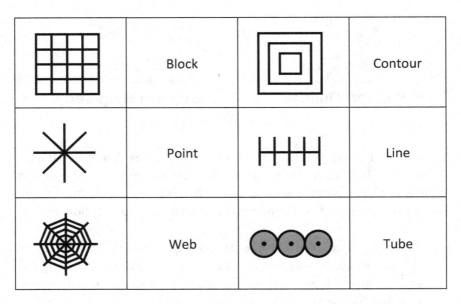

Figure 2.9: Geometries of support structures for metal parts (plan view).

(3) *Means of Support Removal.*

Another aspect of support for consideration is the means of support structure removal, which includes mechanical removal, solvent removal and thermal removal. Support can be mechanically removed by simply breaking away the support structure or with the aid of machining tools. Mechanical removal of support structure usually requires a manual operation that imposes risks of damaging the printed part. On the other hand, a soluble support structure uses different material from the 3D printed part, in which the support structure can be cleaned away by the usage of cleaning solvent. Examples of cleaning solvent include *WaterWorks*^{TM} and *EcoWorks*^{TM} as developed by Stratasys Ltd., which is suitable for the printing of ABS, PC and Nylon parts. Alternatively, thermal energy can also be a means to remove support structure, in which one of the examples can be seen in the RFP technology. The 3D printed part that has soluble or thermally removable support structure is capable of having intricate geometry, internal cavities and better surface finishing. It also requires minimal human contact with the 3D printed part, thus enabling the automation of post-processing, which facilitates the production of multiple small, intricate parts.

2.4 Data Conversion and Transmission

Currently, for most AM systems, the solid or surface model to be built is converted into a format dubbed the .STL file format. This format originates from 3D Systems, which pioneered the stereolithography system. The STL file format approximates the surfaces of the model using tiny triangles. Highly curved surfaces must employ many more triangles, which means that STL files for curved parts can be very large. The STL file format will be discussed in detail in Chap. 6.

Almost, if not all, major CAD-CAM vendors supply the CAD-STL interface. Since 1990, all major CAD-CAM vendors have developed and integrated this interface into their systems.

This conversion step is probably the simplest and shortest of the entire process chain. However, for a highly complex model coupled with an extremely low-performance workstation or PC, the conversion can take several hours. Otherwise, the conversion to STL file should only take several minutes. Where necessary, supports are also converted to a separate STL file. Supports can alternatively be created or modified in the next step by third party software, which allows verification and modifications of models and supports.

A new standard, called the ISO/ASTM 52915:2013 and based primarily on the STL format, is the additive manufacturing file format (AMF) Version 1.1. This new standard is issued jointly under a collaboration between ASTM and ISO in 2013, which answers the growing need of a standard interchangeable file format that can provide detailed properties of the target product. This is discussed in detail in Chap. 6.

The transmission step is also fairly straightforward. The purpose of this step is to transfer the STL files stored in the CAD computer to the computer of the AM system. It is typical that the CAD computer and the AM system are situated in different locations. The CAD computer, being a design tool, is typically located in a design office. The AM system, on the other hand, is a process or production machine and is usually located on the shop floor. More recently, however, there are several AM systems, usually the Concept Modellers (or simply called 3D printers), which can be located in the design office. Data transmission via agreed data formats such as STL or IGES may be carried out through a thumb drive, electronic mail (e-mail) or local area network (LAN). No validation of the quality of the STL files is carried out at this stage.

2.5 Checking and Preparing

The computer term, *garbage in garbage out*, is also applicable to AM. Many first-time users are frustrated at this step to discover that their STL files are faulty. However, more often than not, it is due to both the errors of CAD models and the non-robustness of the CAD-STL interface. Unfortunately, today's CAD models, whose quality is dependent on CAD systems, human operators and post-processes, are still afflicted with a wide spectrum of problems, including the generation of unwanted shell-punctures (i.e., holes, gaps and cracks). These problems, if not rectified, will result in the frequent failure of applications downstream.

These problems are discussed in detail in the first few sections of Chap. 6.

Many of the CAD model errors are corrected by human operators assisted by specialised software such as Magics, a software developed by Materialise, Belgium.[6] Magics software enables users to import a wide variety of CAD formats and to export STL files ready for AM, tooling and manufacturing. Its applications include repairing and optimising 3D models; analysing parts; making process-related design changes on the STL files; documentation; and production planning.[7-9] The manual repair process is, however, still very tedious and time-consuming, especially considering the great number of geometric entities (e.g., triangular facets) encountered in a CAD model. The types of errors and their possible solutions are discussed in Chap. 6.

Once the STL files are verified to be error-free, the AM system's computer analyses the STL files that define the model to be fabricated and slices the model into cross-sections. The cross-sections are systematically recreated through the solidification of liquids, binding of powders or fusing of solids, to form a 3D model.

In SLA, for example, each output file is sliced into cross-sections, between 0.12 mm (minimum) and 0.50 mm (maximum) in thickness. Generally, the model is sliced into the thinnest layer (approximately 0.12 mm) as they have to be very accurate. The supports can be created using coarser settings. An internal cross hatch structure is generated between the inner and the outer surface boundaries of the part. This serves to hold up the walls and entrap liquid that is later solidified using UV light.

Preparing building parameters for positioning and stepwise manufacturing in the light of many available possibilities can be difficult if not accompanied by proper documentation. These possibilities include determination of the geometrical objects, building orientation, spatial assortments, arrangement with other parts, necessary support structures and slice parameters. As in the case of SLA, they also include the determination of technological parameters such as cure depth, laser power and other physical parameters. It means that user-friendly software for ease of use and handling, user support in terms of user manuals, dialogue mode and online graphical aids will be very helpful to users of the AM system.

Many vendors are continually working to improve their systems and their operation software. For example, 3D Systems' Buildstation 5.5 software[10] enables users to simplify the process of setting parameters for the SLA. In early SLA systems, parameters (such as the location in the 250 mm × 250 mm box and the various cure depths) had to be set manually. This was very tedious for up to 12 parameters had to be keyed in. These parameters are shown in Table 2.1.

However, the job is now made simpler with the introduction of default values that can be altered to other specific values. These values can be easily retrieved for use in other models. This software also allows the user to orientate and move the model such that the whole model is in the positive axis' region (the SLA uses only positive numbers for calculations). Thus the original CAD design model can also be in "negative" regions when converting to STL format.

Table 2.1: Parameters used in the SLA process.

1.	X-Y shrink
2.	Z shrink
3.	Number of copies
4.	Multi-part spacing
5.	Range manager (add, delete, etc.)
6.	Recoating
7.	Slice output scale
8.	Resolution
9.	Layer thickness
10.	X-Y hatch-spacing or 60/120 hatch-spacing
11.	Skin fill spacing (X, Y)
12.	Minimum hatch intersecting angle

2.6 Building

For most AM systems, this step is fully automated. Thus, it is usual for operators to leave the machine on to build part(s) overnight. The building process may take up to several hours depending on the size and number of parts required. The number of identical parts that can be built is subjected to the overall build size and constrained by the build volume of the AM system. Most AM systems come with user alert systems that inform the users remotely via electronic communication, for example, cellular phone, once the building of the part is complete.

2.7 Post-Processing

The final task in the process chain is post-processing. At this stage, generally some manual operations are necessary. As a result, the danger of damaging a part is particularly high. Therefore, the operator for this last process step has a high responsibility for the successful realisation of the entire process. The necessary post-processing tasks for some major AM systems are shown in Table 2.2.

The cleaning task refers to the removal of excess materials which may have remained on the part. Thus, for SLA parts, this refers to excess resin residing in entrapped portions such as the blind hole of a part, as well as the removal of supports. Similarly, for SLS part, excess powder has to be removed. Likewise for selective laser melting (SLM), excess metal powder and built supports have to be removed mechanically.

Table 2.2: Essential post-processing tasks for different AM processes.

Additive Manufacturing Technologies			
Post-processing tasks	SLA	SLS	SLM	FDM
1. Cleaning	√	√	√	X
2. Postcuring	√	X	X	X
3. Finishing	√	√	√	√

√ = required; X = not required; SLA — stereolithography apparatus; SLS — selective laser sintering; SLM — selective laser melting; FDM — fused deposition modelling.

As shown in Table 2.2, the SLA post-processing procedures require the highest number of tasks. More importantly, for safety reasons, specific recommendations for post-processing tasks have to be prepared, especially for the cleaning of SLA parts. It was reported by Peiffer that accuracy is related to post-treatment process.[11] Specifically, Peiffer referred to the swelling of SLA-built parts with the use of cleaning solvents. Parts are typically cleaned with solvents to remove unreacted photosensitive resin. Depending upon the "build style" and the extent of crosslinking in the resin, the part can be distorted during the cleaning process. This effect was particularly pronounced with the more open "build styles" and aggressive solvents. With the "build styles" approaching a solid fill and more solvent–resistant materials, damage with the cleaning solvent has been minimised. With newer cleaning solvents like tripropylene glycol monomethyl ether (TPM), which was introduced by 3D Systems, part damage due to the cleaning solvent can be reduced or even eliminated.[11]

For reasons which will be discussed in Chap. 3, SLA parts are built with pockets of liquid embedded within the part. Therefore, post-curing is required. All other non-liquid AM methods do not need to undergo this task.

Finishing refers to secondary processes such as sanding and painting, which are used primarily to improve the surface finish or aesthetic appearance of the part. It also includes additional machining processes such as drilling, tapping and milling to add necessary features to the parts.

References

1. Burns, M (1994). Research notes. *Rapid Prototyping Report*, 4(3), 3–6.

2. Burns, M (1993). *Automated Fabrication*. New Jersey: PTR Prentice Hall.

3. Kochan, D, CK Chua and ZH Du (1999). Rapid prototyping issues in the 21st century. *Computers in Industry*, 39(1), 3–10.

4. Järvinen, J-P, V Matilainen, X Li, H Piili, A Salminen, I Mäkelä and O Nyrhilä (2014). Characterization of effect of support structures in laser additive manufacturing of stainless steel. *Physics Procedia*, 56, 72–81.

5. Krol, TA, MF Zäh and C Seidel (2012). Optimization of supports in metal-based additive manufacturing by means of finite element models. In *Proc. of the 22. Solid Freeform Fabrication Conference*, pp. 707–718. Austin, Texas.

6. Materialise, NV (2008). Magics 3.01 Software for the RP&M Professional. http://www.materialise.com/materialise/view/en/92074-Magics.html.

7. Chua, CK, JGK Gan and T Mei (1997). Interface between CAD and rapid prototyping systems part I: a study of existing interfaces. *International Journal of Advanced Manufacturing Technology*, 13(8), 566–570.

8. Chua, CK, JGK Gan and T Mei (1997). Interface between CAD and rapid prototyping systems part II: LMI — an improved interface. *International Journal of Advanced Manufacturing Technology*, 13(8), 571–576.

9. Gan, JGK, CK Chua and T Mei (1999). Development of new rapid prototyping interface. *Computers in Industry*, 39(1), 61–70.

10. 3D Systems, SLA System Software (September 2008). Retrieved from http://www.3dsystems.com/products/software/3d_lightyear/index.asp.

11. Peiffer, RW (1993). The laser stereolithography process — photosensitive materials and accuracy. *Proceedings of the First International User Congress on Solid Freeform Manufacturing*. Germany.

Problems

1. What are the three types of automated fabricators? Describe them and give two examples of each.

2. Each one of the following manufacturing processes/methods in Table 2.3 belongs to one of three basic types of fabricators. Tick [√] under the column if you think it belongs to that category. If you think that it is a hybrid machine, you may tick more than one category.

Table 2.3: Manufacturing processes/methods.

S/No	Manufacturing Process	Subtractive	Additive	Formative
1.	Press-working			
2.	SLS[a]			
3.	Plastic Injection Moulding			
4.	CNC Nibbling			
5.	CNC CMM[a]			
6.	SLM[a]			

[a]For a list of abbreviations used, refer to the front part of the book.

3. Describe the five steps involved in a general AM process chain. Which steps do you think are likely to be iterated?

4. After 3D geometric modelling, a user can either make a part through NC programming or through AM. What are the basic differences between NC programming and AM in terms of the CAD model?

5. What are the main purposes of the support structure in AM technology? Do you think that all AM technologies require the aid of support structure? Why?

6. STL files are problematic. Is it fair to make this statement? Discuss.

7. Preparing for building appears to be fairly sophisticated. In the case of SLA, what are some of the considerations and parameters involved?

8. Distinguish between cleaning, post-curing and finishing which are the various tasks of post-processing. Name two AM processes that do not require post-curing and one that does not require cleaning.

9. Which step in the entire process chain is, in your opinion, the shortest? Most tedious? Most automated? Support your choice.

CHAPTER 3

LIQUID-BASED ADDITIVE MANUFACTURING SYSTEMS

Most liquid-based additive manufacturing (AM) systems build parts in a vat of photocurable liquid resin, an organic resin that cures or solidifies under the effect of exposure to light, usually in the UV range. The light cures the resin near the surface, forming a thin hardened layer. Once the complete layer of the part is formed, it is lowered by an elevation control system to allow the next layer of resin to be coated and similarly formed over it. This continues until the entire part is completed. The vat can then be drained and the part removed for further processing, if necessary. There are variations to this technique by the various vendors and they are dependent on the type of light or laser, method of scanning or exposure, type of liquid resin and type of elevation and optical system used. Another method is by jetting drops of liquid photopolymer onto a build tray via a printhead, akin to inkjet printing and curing them with UV light. Again, there are variations to the technique depending on the types of resin, exposure, elevation and so on.

3.1 3D Systems' Stereolithography Apparatus

3.1.1 *Company*

3D Systems was founded in 1986 by inventor Charles W. Hull and entrepreneur Raymond S. Freed. Among all the commercial AM systems, the Stereolithography Apparatus, or SLA® as it is commonly called, is the pioneer with its first commercial system marketed in 1988. The company has grown significantly through increased sales and acquisitions, most notably of EOS GmbH's stereolithography business in

1997 and DTM Corp., the maker of the selective laser sintering (SLS) System in 2001. By 2007, 3D Systems had grown into a global company that delivered advanced rapid prototyping solutions to every major market around the world. It has a global portfolio of nearly 400 U.S. and foreign patents, with additional patents filed or pending in the United States and several other major industrialised countries. 3D Systems Inc. is headquartered in 333 Three D Systems Circle Rock Hill, SC 29730 USA.

3.1.2 *Products*

3D Systems produces a wide range of AM machines to cater to various part sizes and throughput. There are several models available, including those in the series of *ProJet*® *6000* SD, HD and MP, *ProJet*® *7000* SD, HD, *ProX*™ 850 and 950. The *ProJet*® *6000* and *ProJet*® *7000* series are low-cost machines that can produce SLA parts.[1,2] The *ProJet* printers have two different sizes, three high definition print configurations and a wide range of *VisiJet*® SL print materials such as Tough, Flexible, Black, Clear, HiTemp, Impact and Jewel.

For larger build envelopes, the *ProX*™ *800* and *950* (see Fig. 3.1) are available. These machines are used to create casting patterns, moulds, end-use parts and functional prototypes. These products' surface smoothness, feature resolution, edge definition and tolerances are enhanced. The *ProX*™ *950* has single-part durability allowing for product build up to 1.5 m wide in one piece without any assembly required. Furthermore, it also uses material efficiently, whereby all unused material will remain in the system resulting in minimal waste. Specifications of these machines are summarised in Tables 3.1(a) and (b).

These machines use one-component, photocurable liquid resins as the material for building. There are several grades of resins available and usage is dependent on the machine's laser and the mechanical requirements of the part. Specific details on the correct type of resins to

Figure 3.1: *ProX*™ 950 printer.
Courtesy of 3D Systems

Table 3.1(a): Summary specifications of the Viper *ProJet*® *6000* SD, HD and MP, *ProJet*® *7000* SD, HD and MP machines.

Models	*ProJet*® *6000 SD*	*ProJet*® *6000 HD*	*ProJet*® *6000 MP*	*ProJet*® *7000 SD*	*ProJet*® *7000 HD*	*ProJet*® *7000 MP*
Net Build Volume (X × Y × Z)	Tall 10 × 10 × 10 in (250 × 250 × 250 mm) Medium 10 × 10 × 5 in (250 × 250 × 125 mm) Short 10 × 10 × 2 in (250 × 250 × 50 mm)			Tall 15 × 15 × 10 in (380 × 380 × 250 mm) Medium n/a Short 15 × 15 × 2 in (380 × 380 × 50 mm)		
Resolution	HD–0.125 mm layers UHD–0.100 mm layers	HD–0.125 mm layers UHD–0.100 mm layers XHD–0.050 mm layers	HD–0.125 mm layers UHD–0.100 mm layers XHD–0.050 mm layers	HD–0.125 mm layers UHD–0.100 mm layers	HD–0.125 mm layers UHD–0.100 mm layers XHD–0.050 mm layers	HD–0.125 mm layers UHD–0.100 mm layers XHD–0.050 mm layers
Accuracy	0.001–0.002 inch (0.025–0.05 mm) per inch of part dimension Accuracy may vary depending on build parameters, part geometry and size, part orientation and post-processing methods					

Courtesy of 3D Systems

Table 3.1(a): (*Continued*) Summary specifications of the Viper *ProJet® 6000* SD, HD and MP, *ProJet® 7000* SD, HD and MP machines.

Models	*ProJet® 6000 SD*	*ProJet® 6000 HD*	*ProJet® 6000 MP*	*ProJet® 7000 SD*	*ProJet® 7000 HD*	*ProJet® 7000 MP*
Materials	*VisiJet®* SL Flex *VisiJet®* SL Tough *VisiJet®* SL Clear *VisiJet®* SL Black *VisiJet®* SL Impact *VisiJet®* SL HiTemp	*VisiJet®* SL Flex *VisiJet®* SL Tough *VisiJet®* SL Clear *VisiJet®* SL Black *VisiJet®* SL Impact *VisiJet®* SL HiTemp *VisiJet®* SL Jewel	*VisiJet®* SL Flex *VisiJet®*SL Tough *VisiJet®* SL Clear *VisiJet®* SL Black *VisiJet®* SL Impact *VisiJet®* SL HiTemp *VisiJet®* SL e-Stone™ *VisiJet®* SL Jewel	*VisiJet®* SL Flex *VisiJet®* SL Tough *VisiJet®* SL Clear *VisiJet®* SL Black *VisiJet®* SL Impact *VisiJet®* SL HiTemp	*VisiJet®* SL Flex *VisiJet®* SL Tough *VisiJet®* SL Clear *VisiJet®* SL Black *VisiJet®* SL Impact *VisiJet®* SL HiTemp *VisiJet®* SL Jewel	*VisiJet®* SL Flex *VisiJet®*SL Tough *VisiJet®* SL Clear *VisiJet®* SL Black *VisiJet®* SL Impact *VisiJet®* SL HiTemp *VisiJet®* SL e-Stone™ *VisiJet®* SL Jewel
Material Packaging	Material in clean no drip 2.0 litre cartridges System auto fills print tray between builds					
Electrical	100–240 VAC, 50/60 Hz, single-phase, 750 W					
Dimensions (*W × D × H*)	3D Printer Crated 66 × 35 × 79 in (1676 × 889 × 2006 mm) 3D Printer Uncrated 31 × 29 × 72 in (787 × 737 × 1829 mm)			3D Printer Crated 73.5 × 38.5 × 81.5 in (1860 × 982 × 2070 mm) 3D Printer Uncrated 39.0 × 34.0 × 72 in (984 × 854 × 1829 mm)		
Weight	3D Printer Crated 600 lbs (272 kg) 3D Printer Uncrated 400 lbs (181 kg)			3D Printer Crated 800 lbs (363 kg) 3D Printer Uncrated 600 lbs (272 kg)		
3D Manage Software	Easy build job setup, submission and job queue management Automatic part placement and build optimisation tools Part stacking and nesting capability Extensive part editing tools Automatic support generation Job statistics reporting					
MP Auto Software	Automation utility for rapid manufacturing applications. Included only with the *ProJet* 6000 MP			Automation utility for rapid manufacturing applications. Included only with the *ProJet* 7000 MP		
Network Compatibility	Network ready with 10/100 Ethernet interface 4 MB					

Table 3.1(a): (*Continued*) Summary specifications of the Viper *ProJet® 6000* SD, HD and MP, *ProJet® 7000* SD, HD and MP machines.

Models	ProJet® 6000 SD	ProJet® 6000 HD	ProJet® 6000 MP	ProJet® 7000 SD	ProJet® 7000 HD	ProJet® 7000 MP
3D Manage Hardware Recommendation	Core 2 Duo 1.8 GHz with 4 GB RAM (OpenGL support 128 Mb video RAM)					
3D Manage Operating System	Windows XP Professional, Windows Vista, Windows 7					
Input Data File Formats Supported	STL and SLC					
Operating Temperature Range	64–82°F (18–28°C)					
Noise	<65 dBa estimated					
Operational Accessories	UV Curing Units, Parts Washer and Right Height Table, *ProJet®* Cart Station			UV Curing Units, *ProJet®* Cart Station		

Table 3.1(b): Summary specifications of the *ProX™* 800 and *ProX™* 950 machines.

Models	ProX™ 800	ProX™ 950
Net Build Volume (xyz)		
Full	650 × 750 × 550 mm (25.6 × 29.5 × 21.65 in); 414 I (109.3 U.S. gal)	1500 × 750 × 550 mm (59 × 30 × 22 in); 935 I (247 U.S. gal)
Half	650 × 750 × 275 mm (25.6 × 29.5 × 10.8 in); 272 I (71.9 U.S. gal)	n/a
Short	650 × 750 × 50 mm (25.6 × 29.5 × 1.97 in); 95 I (25.09 U.S. gal)	n/a
Resolution	0.00127 mm (0.0005 in) laser spot location resolution	
Accuracy	0.025–0.05 mm per 25.4 mm (0.001–0.002 in per inch) of part dimension. Accuracy may vary depending on build parameters, part geometry and size, part orientation and post-processing methods	
Materials	Builds with broadest range of 3D printing materials with exceptional mechanical properties. See www.3dsystems.com for available materials	
Material Packaging	Material in clean, no-drip 10 kg cartridges. System auto fills print tray between builds.	

Courtesy of 3D Systems

Table 3.1(b): (*Continued*) Summary specifications of the *ProX*™ 800 and *ProX*™ 950 machines.

Models	*ProX*™ 800	*ProX*™ 950
Electrical Requirements	200–240 VAC 50/60 Hz, single-phase, 30 amps	200–240 VAC 50/60 Hz, single-phase, 50 amps
Dimensions		
3D Printer Crated	190 × 163 × 248 cm (75 × 64 × 98 in)	242 × 173 × 254 cm (95 × 68 × 100 in)
3D Printer Uncrated	137 × 160 × 226 cm (50 × 63 × 89 in)	220 × 160 × 226 cm (87 × 63 × 89 in)
Weight		
3D Printer Crated	1134 kg (2500 lbs)	1724 kg (3800 lbs)
3D Printer Uncrated	1724 kg (3800 lbs) not including MDM	1951 kg (4300 lbs) not including MDM
Print3DPro and *3DManage*™ Software	Easy build job setup, submission and job queue management Automatic part placement and build optimization tools Part nesting capability Extensive part editing tools Automatic support generation Job statistics reporting	
Network Compatibility	Ethernet, IEEE 802.3 using TCP/IP and NFS	
3D Manage Hardware Recommendation	I5, 2.3 GHz with 8 GB RAM (Open GL support 1 GB video RAM)	
Software Operating System	*Windows*® 7 and newer	
Input data File Formats Supported	.STL and .SLC	
Operating Temperature range	20–26°C (68–79°F)	
Noise	Not to exceed 70 dBA	
Accessories	Interchangeable quick change material deliverable modules (MDMs) with integrated elevator and removable applicator Manual offload cart ProCure™ 750 or 1500 UV Finisher	

be used are available from the manufacturer. The other main consumable used by these machines is the cleaning solvent which is required to clean the part of any residual resin after the building of the part is completed on the machine.

3.1.3 *Process*

3D Systems' stereolithography process creates three-dimensional (3D) plastic objects directly from computer-aided design (CAD) data. The process begins with the vat filled with the photocurable liquid resin and the elevator table set just below the surface of the liquid resin (see Fig. 3.2). The *3DManage*™ loads a 3D CAD solid model file into the system. 3DPrint™ software takes the output files from the *3DManage*™ software, and then provides control of the build via the attached controller PC on the SLA system. Supports are designed to stabilise the part during building. The translator converts the CAD data into an STL file. The control unit slices the model and support into a series of cross sections from 0.025 to 0.5 mm (0.001–0.020 in.) thick. The computer-controlled optical scanning system then directs and focuses the laser beam so that it solidifies a two-dimensional (2D) cross section corresponding to the slice on the surface of the photocurable liquid resin to a depth greater than one-layer thickness. The elevator table then lowers enough to cover the solid polymer with another layer of the liquid resin. A levelling wiper or vacuum blade moves across the surfaces to

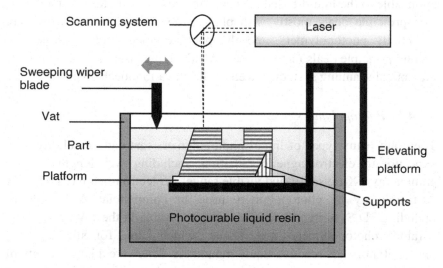

Figure 3.2: Schematic of SLA process.

recoat the next layer of resin on the surface. The laser then draws the next layer. This process continues building the part from bottom up, until the system completes the part. The part is then raised out of the vat and cleaned of excess polymer.

3.1.4 *Principle*

The SLA process is based fundamentally on the following principles[3]:

(1) Parts are built from a photocurable liquid resin that cures when exposed to a laser beam (basically, undergoing the photopolymerisation process), which scans across the surface of the resin.
(2) The building is done layer by layer, each layer being scanned by the optical scanning system and controlled by an elevation mechanism which lowers at the completion of each layer.

These two principles will be discussed briefly in this section to lay the foundation for the understanding of AM processes. They are mostly applicable to the liquid-based AM systems described in this chapter. This first principle deals mostly with photocurable liquid resins, which are essentially photopolymers, and the photopolymerisation process. The second principle deals mainly with CAD data, the laser and the control of the optical scanning system as well as the elevation mechanism.

3.1.4.1 *Photopolymers*

There are many types of liquid photopolymers that can be solidified by exposure to electromagnetic radiation, including wavelengths in the gamma rays, X-rays, UV and visible range, or electron-beam (EB).[4,5] The vast majority of photopolymers used in commercial AM systems, including 3D Systems' SLA machines are curable in the UV range. UV-curable photopolymers are resins which are formulated from photoinitiators and reactive liquid monomers. There are a large variety of them and some may contain fillers and other chemical modifiers to meet specified chemical and mechanical requirements.[6] The process through which photopolymers are cured is referred to as the photopolymerisation process.

3.1.4.2 *Photopolymerisation*

Loosely defined, polymerisation is the process of linking small molecules (known as monomers) into chain-like larger molecules (known as polymers). When the chain-like polymers are linked further to one another, a cross-linked polymer is said to be formed. Photopolymerisation is polymerisation initiated by a photochemical process whereby the starting point is usually the induction of energy from an appropriate radiation source.[7]

Polymerisation of photopolymers is normally an energetically favourable or exothermic reaction. However, in most cases, the formulation of photopolymer can be stabilised to remain unreacted at ambient temperature. A catalyst is required for polymerisation to take place at a reasonable rate. This catalyst is usually a free radical which may be generated either thermally or photochemically. The source of a photochemically generated radical is a photoinitiator, which reacts with an actinic photon to produce the radicals that catalyse the polymerisation process.

The free radical photopolymerisation process is schematically presented in Fig. 3.3.[8] Photoinitiator molecules, P_i, which are mixed with the monomers, M, are exposed to a UV source of actinic photons, with energy of $h\nu$, where h is the Planck constant and ν is the frequency of the radiation. The photoinitiators absorb some of the photons and are in an excited state. Some of these are converted into reactive initiator molecules, P•, after undergoing several complex chemical energy transformation steps. These molecules then react with a monomer molecule to form a polymerisation initiating molecule, PM•. This is the chain initiation step. Once activated, additional monomer molecules go on to react in the chain propagation step, forming longer molecules, PMMM• until a chain inhibition process terminates the polymerisation reaction. The longer the reaction is sustained, the higher will be the molecular weight of the resulting polymer. Also, if the monomer molecules have three or more reactive chemical groups, the resulting polymer will be cross-linked, and this will generate an insoluble continuous network of molecules.

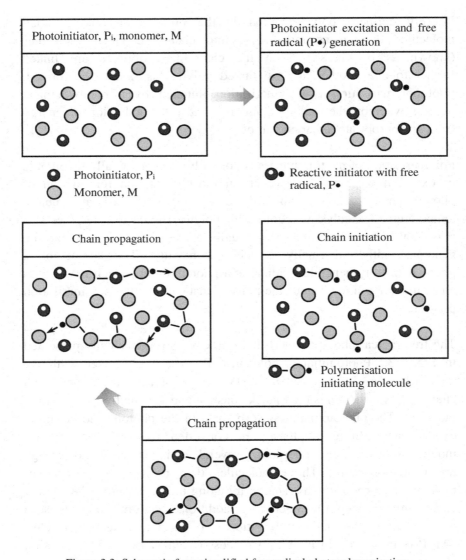

Figure 3.3: Schematic for a simplified free radical photopolymerisation.

During polymerisation, it is important that the polymers are sufficiently cross-linked so that the polymerised molecules do not re-dissolve back into the liquid monomers. The photopolymerised molecules must also possess sufficient strength to remain structurally sound while the cured resin is subjected to various forces during recoating.

While free radical photopolymerisation is well established and yields polymers that are acrylate-based, there is another newer "chemistry" known as cationic photopolymerisation.[9] It relies on cationic initiators, usually iodinium or sulfonium salts, to start polymerisation. Commercially available cationic monomers include epoxies, the most versatile of cationally polymerisable monomers and vinyl ethers. Cationic resins are attractive as prototype materials as they have better physical and mechanical properties. However, the process may require higher exposure time or a higher power laser.

3.1.4.3 *Layering technology, laser and laser scanning*

Almost all AM systems use layering technology in the creation of prototype parts. The basic principle is the availability of computer software to slice a CAD model into layers and reproduce it in an "output" device like a laser scanning system. The layer thickness is controlled by a precision elevation mechanism. It will correspond directly to the slice thickness of the computer model and the cured thickness of the resin. The limiting aspect of the AM system tends to be the curing thickness rather than the resolution of the elevation mechanism.

The important component of the building process is the laser and its optical scanning system. The key strength of the SLA is its ability to rapidly direct focused radiation of appropriate power and wavelength onto the surface of the liquid photopolymer resin, forming patterns of solidified photopolymer according to the cross-sectional data generated by the computer.[10] In the SLA, a laser beam with a specified power and wavelength is sent through a beam expanding telescope to fill the optical aperture of a pair of cross axis, galvanometer driven and beam scanning mirrors. These form the optical scanning system of the SLA. The beam comes to a focus on the surface of a liquid photopolymer, curing a predetermined depth of the resin after a controlled time of exposure (inversely proportional to the laser scanning speed).

The solidification of the liquid resin depends on the energy per unit area (or "exposure") deposited during the motion of the focused spot on the surface of the photopolymer. There is a threshold exposure that must be exceeded for the photopolymer to solidify.

To maintain accuracy and consistency during part building using the SLA, the cure depth and the cured line width must be controlled. As such, accurate exposure and focused spot size become essential.

Parameters which influence performance and functionality of the parts are the physical and chemical properties of resin, speed and resolution of the optical scanning system, the power, wavelength and type of the laser used, the spot size of the laser, the recoating system and the post-curing process.

3.1.5 *Strengths and weaknesses*

The main strengths of the SLA are:

(1) *Round the clock operation.* The SLA can be used continuously and unattended round the clock.
(2) *Build volumes.* The different SLA machines have build volumes ranging from small (250 × 250 × 250 mm) to large (1500 × 750 × 550 mm) to suit the needs of different users.
(3) *Good accuracy.* The SLA has good accuracy and can thus be used for many application areas.
(4) *Surface finish.* The SLA can obtain one of the best surface finishes among AM technologies.
(5) *Wide range of materials.* There is a wide range of materials, from general-purpose materials to specialty materials for specific applications.

The main weaknesses of the SLA are:

(6) *Requires support structures.* Structures that have overhangs and undercuts must have supports that are designed and fabricated together with the main structure.
(7) *Requires post-processing.* Post-processing includes removal of supports and other unwanted materials, which is tedious, time consuming and can damage the model.
(8) *Requires post-curing.* Post-curing may be needed to cure the object completely and ensure the integrity of the structure.

3.1.6 *Applications*

The SLA technology provides manufacturers cost justifiable methods for reducing time to market, lowering product development costs, gaining greater control of their design process and improving product design. The range of applications includes:

(1) Models for conceptualisation, packaging and presentation.
(2) Prototypes for design, analysis, verification and functional testing.
(3) Parts for prototype tooling and low-volume production tooling.
(4) Patterns for investment casting, sand casting and moulding.
(5) Tools for fixture and tooling design, and production tooling.

Software developed to support these applications includes *QuickCast*™, a software tool which is used in the investment casting industry. *QuickCast*™ enables highly accurate resin patterns that are specifically used as an expendable pattern to form a ceramic mould to be created. The expendable pattern is subsequently burnt out. The standard process uses an expendable wax pattern which must be cast in a tool. *QuickCast*™ eliminates the need for the tooling use to make the expendable patterns. *QuickCast*™ produces parts which have a hard thin outer shell and contain a honeycomb-like structure inside, allowing the pattern to collapse when heated instead of expanding, which would crack the shell.

3.1.7 *Research and development*

To stand as a leading company in the competitive market and to improve customers' needs, 3D Systems' research has made great improvement not only in developing new materials and applications, but also in software applications.

At the time of writing, 3D Systems announced they are collaborating with the White House, UI Labs, the Department of Defence and other industries and organisations to deliver their latest *Geomagic®* perceptual design, manufacturing tools and inspection products to allow unparalleled access to the most advanced design to manufacturing digital thread for the manufacturing industry.[11]

The Digital Lab for Manufacturing is an applied research institute geared to transform the manufacturing chain through development, demonstration and deployment of digital manufacturing technologies.

3D Systems has also recently bought Xerox's colour printing technology and their development labs where Xerox's engineering and development teams for product design and materials science will be integrated into 3D Systems' portfolio.[12] This acquisition would bolster their technology and product development to allow 3D Systems to bring affordable and high performance new products to market.

3.2 **Stratasys'** *PolyJet*

3.2.1 *Company*

Stratasys Inc. was founded in 1989 in Delaware and developed the company's AM systems based on Fused Deposition Modelling (FDM) technology. In December 2012, Stratasys Inc. merged with Objet. Objet was founded in 1998 and has established itself as the leading platform for high-resolution 3D printing. Using its patented and market-proven *PolyJet*™ inkjet-head technology, it is able to print out the most complex

3D models with exceptionally high quality. As a company, Stratasys Inc. is located at 7665 Commerce Way, Eden Prairie, MN 55344-2080, USA.

3.2.2 *Products*

Stratasys manufactures 3D printing equipment and materials that create physical objects based on digital data. Its systems range from affordable desktop 3D printers to large, advanced 3D production systems. All Stratasys 3D printers build parts layer-by-layer.

3.2.2.1 *Production Series — Precision 3D Printers*

Stratasys' Production Series 3D printers aim to produce parts without high expense and longtime duration. The 3D Production Systems are driven by PolyJet technology work by jetting state-of-the-art photopolymer materials in ultra-thin layers onto a build tray, layer by layer, until the part is complete. Under Production Series, there are two categories of 3D printers — Precision and Performance 3D printers. Precision 3D printers set the industry standard for making final products. On the other hand, Performance 3D printers are further discussed in Section 4.1.

Production series — Precision Series consists of *Connex3*, *Objet1000 Plus* and *Stratasys J750*. *Connex3* systems can deliver high-resolution prototypes with varies material properties within a single print job. *Objet1000 Plus* has very high throughput and amplifies productivity without sacrificing accuracy. The *Stratasys J750* 3D printer (see Fig. 3.4), the newest model in the *PolyJet*™ series, is more sophisticated in terms of automatic colour mapping, easy material selection and finer layers and faster 3D printing. Specifications of the Production Series — Precision 3D printers are summarized in Table 3.2.

3.2.2.2 *Materials*

Stratasys Inc. offers a range of AM materials, including clear, rubber-like and biocompatible photopolymers, and tough high performance

Figure 3.4: *Stratasys J750* printer.
Courtesy of Stratasys Ltd.

Table 3.2: Summary specifications of *Connex3*, *Objet1000 Plus* and *Stratasys J750*.

Models	Connex3	Objet1000 Plus	Stratasys J750
Model Materials	Vero family of opaque materials, including various neutral shades and vibrant colours Tango family rubber-like flexible materials Medical: MED610 Digital ABS and Digital ABS2 in ivory and green DurusWhite RGD430 High Temperature RGD525 white Transparent: VeroClear and RGD720	Rigid Opaque: Vero family rubber-like: TangoPlus and TangoBlackPlus Transparent: VeroClear Simulated Polypropylene (Rigur)	Vero family of opaque materials including neutral shades and vibrant colours Tango family of flexible materials Transparent VeroClear and RGD720

Courtesy of Stratasys Inc.

Table 3.2: (*Continued*) Summary specifications of *Connex3*, *Objet1000 Plus* and *Stratasys J750*.

Models	Connex3	Objet1000 Plus	Stratasys J750
Digital Model Materials	Hundreds of composite materials can be manufactured on the fly including: Digital ABS Rubber-like materials in a variety of Shore A values Vibrant blended colors in rigid opaque Translucent colored tints Polypropylene-like materials with improved thermal resistance	Digital ABS and Digital ABS2 Wide range of translucencies Rubber-like blends in a range of Shore A values Simulated Polypropylene materials with improved heat resistance	Unlimited number of composite materials including: • Over 360,000 colours • Digital ABS and Digital ABS2 in ivory and green • Rubber-like materials in a variety of Shore A values • Translucent color tints
Support Material	SUP705 (WaterJet removable) SUP706 (soluble)	FullCure 705 non-toxic gel-like photopolymer support	SUP705 (WaterJet removable)
Net Build Size	Objet350 *Connex3*: 340 × 340 × 200 mm (13.4 × 13.4 × 7.9 in.) Objet500 *Connex3*: 490 × 390 × 200 mm (19.3 × 15.4 × 7.9 in.)	1000 × 800 × 500 mm (39.3 × 31.4 × 19.6 in.)	490 × 390 × 200 mm (19.3 × 15.35 × 7.9 in.)
Layer Thickness	Horizontal build layers down to 16 microns (.0006 in.)	Horizontal build layers as fine as 16 microns (.0006 in.)	Horizontal build layers down to 14 microns (0.00055 in.)
Build Resolution	X-axis: 600 dpi Y-axis: 600 dpi Z-axis: 1600 dpi	X-axis: 300 dpi Y-axis: 300 dpi Z-axis: 1600 dpi	X-axis: 600 dpi Y-axis: 600 dpi Z-axis: 1800 dpi

Table 3.2: (*Continued*) Summary specifications of *Connex3*, *Objet1000 Plus* and *Stratasys J750*.

Models	Connex3	Objet1000 Plus	Stratasys J750
Printing Modes	Digital material (DM): 30-micron (0.001 in.) resolution High quality (HQ): 16-micron (0.0006 in.) resolution High speed: 30-micron (0.001 in.) resolution	DM: 34-micron (0.001 in.) HQ: 16-micron (0.0006 in.) High Speed (HS): 34-micron (0.001 in.)	HS: up to 3 base resins, 27-micron (0.001 in.) resolution HQ: up to 6 base resins, 14-micron (0.00055 in.) resolution High Mix: up to 6 base resins, 27-micron (0.001 in.) resolution
Accuracy	20–85 micron for features below 50 mm; up to 200 micron for full model size (for rigid materials only, depending on geometry, build parameters and model orientation)	Up to 85 microns for features smaller than 50 mm; Up to 600 microns for full model size (for rigid materials only, depending on geometry, build parameters and model orientation)	20–85 microns for features below 50 mm; up to 200 microns for full model size (for rigid materials only)
Workstation Compatibility	Windows 7 and Windows 8	Integrated Windows 7 64 bit and Windows 8	Windows 7 and 8.1 64 bit
Network Connectivity	LAN-TCP/IP		
Size and Weight	1400 × 1260 × 1100 mm (55.1 × 49.6 × 43.3 in.) 430 kg (948 lbs)	Height: 1960 mm (77.5 in.) Width: 2868 mm (113 in.) Depth: 2102 mm (83 in.) Weight: 2200 kg (4850 lbs)	3D Printer: 1400 × 1260 × 1100 mm (55.1 × 49.6 × 43.4 in.); 430 kg (948 lbs) Material Cabinet: 670 × 1170 × 640 mm (26.4 × 46.1 × 25.2 in.); 152 kg (335 lbs)

Table 3.2: (*Continued*) Summary specifications of *Connex3*, *Objet1000 Plus* and *Stratasys J750*.

Models	Connex3	Objet1000 Plus	Stratasys J750
Power Requirements	100–120 VAC 50/60 Hz; 13.5 A 200–240 VAC 50/60 Hz; 7 A single phase	230 VAC 50/60 Hz; 8 A single phase	100–120 VAC, 50–60 Hz, 13.5 A, 1 phase 220–240 VAC, 50–60 Hz, 7 A, 1 phase
Operational Environment	Temperature 18–25°C (64–77°F); relative humidity 30%–70% (non-condensing)	Temperature 18–22°C (64.5–71.5°F); relative humidity 30%–70%	Temperature 18–25°C (64–77°F); relative humidity 30%–70% (non-condensing)

thermoplastics. This variety enables customers to maximise their benefits of 3D printing throughout the product development cycle. From fast, affordable concept modelling to detailed, super-realistic functional prototyping, through certification testing and into agile, low-risk production, Stratasys materials help designers and engineers succeed at every stage. PolyJet photopolymers offer a high level of creating all the details and final products. With chameleon-like ability to simulate clear, flexible and rigid materials and engineering plastics, combined with multiple material properties into one model, prototypes can be made as final products quickly.

3.2.3 *Process*

PolyJet 3D printing is similar to inkjet document printing. But instead of jetting drops of ink onto paper, *PolyJet* 3D printers jet layers of liquid photopolymer onto a build tray and cure them with UV light. The layers build up one at a time to create a 3D model or prototype. Fully cured models can be handled and used immediately, without additional post-curing. Along with the selected model materials, the 3D printer also jets a gel-like support material specially designed to uphold overhangs and complicated geometries. It is easily removed by hand and with water.

Stratasys' PolyJet process creates 3D objects with the use of Objet Studio Software. The designer loads the 3D CAD solid model file into the system which is compatible with Windows XP and Windows 2000. The Objet Studio will convert the CAD data into an STL or SLC file. The designer will also have to set the orientation arrangement of the designed part on the build tray.

Before the actual building process commences, the designer has to ensure that the build tray and the two types of material cartridges are inserted into the machine. The two types of material cartridges consist of the part material and the supporting material. When the procedure is done, the jetting heads, based on Stratasys' patented PolyJet inkjet technology, will move along the x-axis and lay the first layer of material onto the tray. Depending on the size of the part, the jetting head will move on the y-axis and move on to the x-axis to lay the next layer after the first layer is completed. During the printing process, the jetting head will release the actual amounts of part material and support material. The materials will be immediately cured by the UV light from the jetting head. Whenever the materials are about to be used up, the material cartridges can be easily replaced without interrupting the fabrication process. Once the jetting head cures the first 2D cross section, the build tray will drop by one-layer thickness of 16 μm. The jetting head will repeat the process continuously until the system completes the part. The part will be raised up and can be taken out for post-processing. The support material can then be removed easily by the water jet and the part is complete.

3.2.4 *Principle*

Stratasys' *PolyJet*™ technology creates high-quality models directly from the computerised 3D files. Complex parts are produced with the combination of Stratasys' Studio software and the jetting head.

The process is based on the following principles:

(1) Jetting heads release the required amount of material which shares the same method as the normal inkjet printing method. At the same time when the material is printed on the tray, the material is cured

by the UV light which is integrated with the jetting head. Parts are built layer by layer, from a liquid photopolymer where a similar polymerisation process as described in Section 3.1.4 takes place.

(2) Jetting heads are moved only along the XY-axes and each slice of the building process is the cross section of the parts arranged in the software.

(3) With the completion of a cross-section layer, the build tray will be lowered for the next layer to be laid. The Z-height of the elevator is levelled accurately so that the corresponding cross-section data can be calculated for that layer.

(4) Both the part material and support material will be fully cured when they are exposed to the UV light and most importantly the non-toxic support material can be removed easily by the water jet.

3.2.5 *Strengths and weaknesses*

Stratasys' *PolyJet*™ system has the following strengths[13]:

(1) *High quality.* The *PolyJet*™ can build layers as thin as 16 µm in thickness with accurate details depending on the geometry, part orientation and print size.

(2) *High accuracy.* Precise jetting and build material properties enable fine details and thin walls (600 µm or less depending on the geometry and materials).

(3) *Fast process speed.* Certain AM systems require draining, resin stripping, polishing and others whereas *Eden*™ systems only require an easy wash of the support material which is a key strength.

(4) *Smooth surface finish.* The models built have smooth surface and fine details without any post-processing.

(5) *Wide range of materials.* Stratasys has a range of materials suited for different specifications, ranging from tough acrylic-based polymer, to polypropylene-like plastics (Duruswhite) to the rubber-like Tango materials.

(6) *Easy usage.* The *Eden*™ family utilises a cartridge system for easy replacement of build and support materials. Material cartridges provide an easy method for insertion without having any risk of contact with the materials.

(7) *Single head replacement (SHR) technology.* The *Eden*™ machines' nozzles consist of heads and nozzles. With SHR, these individual nozzles can be replaced instead of replacing the whole unit whenever the need arises.

(8) *Safe and clean process.* Users are not exposed to the liquid resin throughout the modelling process and the photopolymer support is non-toxic. *PolyJet*™ systems can be installed in the office environment without increasing the noise level.

(9) *Combination of the materials.* Based on PolyJet technology, Connex 3D printers from Stratasys are the only AM systems that can combine different 3D printing materials within the same 3D printed model, in the same print job.

The *PolyJet*™ system has the following weaknesses:

(1) *Post-processing.* A water jet is required to wash away the support material used in *PolyJet*™, meaning that water supply must be nearby. This is somewhat a let-down to the claim that the machine is suitable for an office environment. In cases where the parts built are small, thin or delicate, the water jet can damage these parts, so care in post-processing must be exercised.

(2) *Wastage.* The support material which is washed away with water cannot be reused, meaning additional costs are added to the support material.

3.2.6 *Applications*

The applications of Stratasys' systems can be divided into different areas:

(1) *General applications.* Models created by Stratasys' systems can be used for conceptual design presentation, design proofing, engineering testing, integration and fitting, functional analysis, exhibitions and preproduction sales, market research and interprofessional communication.

(2) *Tooling and casting applications.* Parts can also be created for investment casting, direct tooling and rapid tool-free manufacturing

of plastic parts. Also, they can be used to create silicon moulding, aluminium epoxy moulds, very low temperature (VLT) moulding (alternative rubber mould) and vacuum forming.

(3) *Medical imaging.* Diagnostic, surgical,[14] operation and reconstruction planning and custom prosthesis design. Parts built by *PolyJet*™ have outstanding detail and fine features which can make the medical problems more visible for analysis and surgery simulation. Due to its fast building time, prototype models are always built for trauma or tumours. Most importantly, it reduces the surgical risks and provides a communication bridge for the patients.

(4) *Jewellery industry.* Presentation of concept design, actual display, design proof and fitting. Premarket survey and market research can be conducted using these models.

(5) *Packing.* Vacuum forming is an easy method to produce inexpensive parts and it requires a very short time for the part to be formed.

3.2.7 *Research and development*

Stratasys is doing research to develop faster processing, higher performance, higher resolution graphics, smoother and more accurate details. One way of improving the accuracy of the PolyJet procedure can be done by optimising the scaling factor (Fig. 3.5).[15]

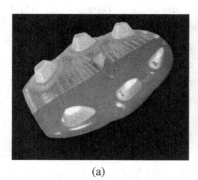
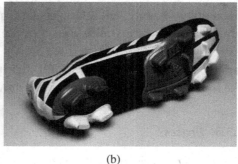

(a)   (b)

Figure 3.5: High quality finishing part that (a) appeared close to a real part and (b) fitted onto a real soccer boot.
Courtesy of Stratasys Inc.

3.3   3D Systems' MultiJet Printing System

3.3.1   *Company*

3D Systems is one of the earliest and largest AM companies in the world. Among its many products, 3D Systems also manufactures 3D-Printing systems meant mainly for the office environment. The MultiJet Printing system (MJP) was formerly known as MultiJet Modelling (MJM), first launched in 1996 as a concept modeller for the office to complement the more sophisticated SLA machines (see Section 3.1) and SLS machines (see Section 5.1). The company's details are found in Section 3.1.1.

3.3.2   *Products*

The *ProJet*™ Series machines are based on MJP technology with thermal material application and UV curing. The newer *ProJet*™ Series machines were launched in 2008. The *ProJet*™ 3500 Printer (see Fig. 3.6) is aimed at offering higher productivity, including the high speed printing mode, and larger high definition prints, for the production of functional plastic parts for product design and manufacturing applications. The *ProJet*™ 5000 printer (see Fig. 3.7) is designed for larger build envelopes to achieve maximum productivity with a net build volume of 553 × 381 × 300 mm. In addition, *ProJet*™ 5500 produces the highest quality, most accurate and toughest multi-material composites based on MJP technology among all the printers produced by 3D System. It simultaneously prints and fuses flexible and rigid material composites layer by layer together. It uses newly developed *VisiJet*® composite family of materials to print all the products. The *ProJet*™ Series machines and their specifications are shown in Table 3.3.

3.3.3   *Process*

The *ProJet*™ Series uses the *ProJet*™ Accelerator Software to set up the "sliced" STL files for the system from the CAD software. The *ProJet*™ Accelerator Software is a powerful software which allows users to verify

Figure 3.6: *ProJet*™ 3500 printer.
Courtesy of 3D Systems

Figure 3.7: *ProJet*™ 5000 printer.
Courtesy of 3D Systems

Table 3.3: Specifications of *ProJet*™ 3500 SD & HD series and *ProJet*™ 5000 and 5500X.

Models	*ProJet*™ 3510 SD	*ProJet*™ 3510 HD	*ProJet*™ 3510 HD*Plus*	*ProJet*™ 3500 HD*Max*	*ProJet*™ 5000	*ProJet*™ 5500X
Printing Modes	HD — High Definition	HD — High Definition - UHD — Ultra-High Definition -	HD — High Definition - UHD — Ultra-High Definition XHD — Xtreme High Definition	HD — High Definition HS — High Speed UHD — Ultra-High Definition XHD — Xtreme High Definition	n/a	n/a
Net Built Volume (xyz) HD Mode HS Mode UHD Mode XHD Mode	11.75 × 7.3 × 8 in. (298 × 185 × 203 mm)	11.75 × 7.3 × 8 in. (298 × 185 × 203 mm) 5 × 7 × 6 in. (127 × 178 × 152 mm)	11.75 × 7.3 × 8 in. (298 × 185 × 203 mm) 5 × 7 × 6 in. (127 × 178 × 152 mm) 5 × 7 × 6 in. (127 × 178 × 152 mm)	11.75 × 7.3 × 8 in. (298 × 185 × 203 mm) 11.75 × 7.3 × 8 in (298 × 185 × 203 mm) 11.75 × 7.3 × 8 in (298 × 185 × 203 mm) 11.75 × 7.3 × 8 in. (298 × 185 × 203 mm)	21 × 15 × 11.8 in. (533 × 381 × 300 mm)	21 × 15 × 11.8 in. (533 × 381 × 300 mm) 21 × 15 × 11.8 in. (533 × 381 × 300 mm)

Table 3.3: (*Continued*) Specifications of *ProJet*™ 3500 SD & HD series and *ProJet*™ 5000 and 5500X.

Models	*ProJet*™ 3510 SD	*ProJet*™ 3510 HD	*ProJet*™ 3510 HD*Plus*	*ProJet*™ 3500 HD*Max*	*ProJet*™ 5000	*ProJet*™ 5500X
Resolution HD Mode HS Mode UHD Mode XHD Mode	375 × 375 × 790 DPI (xyz); 32 μ layers	375 × 375 × 790 DPI (xyz); 32 μ layers 750 × 750 × 890 DPI (xyz); 29 μ layers	375 × 375 × 790 DPI (xyz); 32 μ layers 750 × 750 × 890 DPI (xyz); 29 μ layers 750 × 750 × 1600 DPI (xyz); 16 μ layers	375 × 375 × 790 DPI (xyz); 32 μ layers 375 × 375 × 790 DPI (xyz); 32 μ layers 750 × 750 × 890 DPI (xyz); 29 μ layers 750 × 750 × 1600 DPI (xyz); 16 μ layers	375 × 375 × 790 DPI (xyz); 32 μ layers 375 × 375 × 395 DPI; 64 μ layers 750 × 750 × 890 DPI (xyz); 32 μ layers	375 × 375 × 790 DPI (xyz); 32 μ layers 750 × 750 × 890 DPI (xyz); 29 μ layers
Accuracy (typical)	0.001–0.002 in. (0.025–0.05 mm) per inch of part dimension. Accuracy may vary depending on build parameters, part geometry and size, part orientation and post-processing.					
E-mail Notice Capability	Yes	Yes	Yes	Yes	Yes	–
Tablet/ Smartphone connectivity	Yes	Yes	Yes	Yes	Yes	–
5-Year Printhead Warranty	Optional	Standard	Standard	Standard	Yes	–
Build Materials	*VisiJet* M3 X *VisiJet* M3 Black *VisiJet* M3 Crystal *VisiJet* M3 Proplast *VisiJet* M3 Navy *VisiJet* M3 Techplast	*VisiJet* M3 X *VisiJet* M3 Black *VisiJet* M3 Crystal *VisiJet* M3 Proplast *VisiJet* M3 Navy *VisiJet* M3 Techplast *VisiJet* M3 Procast	*VisiJet* M3 X *VisiJet* M3 Black *VisiJet* M3 Crystal *VisiJet* M3 Proplast *VisiJet* M3 Navy *VisiJet* M3 Techplast *VisiJet* M3 Procast	*VisiJet* M3 X *VisiJet* M3 Black *VisiJet* M3 Crystal *VisiJet* M3 Proplast *VisiJet* M3 Navy *VisiJet* M3 Techplast *VisiJet* M3 Procast	*VisiJet* M5-X *VisiJet* M5-Black *VisiJet* M5-MX	*VisiJet* CR-CL *VisiJet* CR-WT *VisiJet* CF-BK

Table 3.3: (*Continued*) Specifications of *ProJet*™ 3500 SD & HD series and *ProJet*™ 5000 and 5500X.

Models	ProJet™ 3510 SD	ProJet™ 3510 HD	ProJet™ 3510 HD*Plus*	ProJet™ 3500 HD*Max*	ProJet™ 5000	ProJet™ 5500X
Support Material	*VisiJet* S300	*VisiJet* S300	*VisiJet* S300	*VisiJet* S300	*VisiJet* S300	*VisiJet* S500
Material Packaging Build and Support Materials	In clean 4.41 lbs (2 kg) bottles (machine holds up to 2 with autoswitching)				In clean 2.0 kg cartridges. Printer can hold up to eight cartridges with additional material bays (optional)	Build materials in clean 2.0 kg cartridges and support material in clean 1.75 kg cartridges (printer holds 4 build and 4 support cartridges with auto-switching)
Electrical	100–127 VAC, 50/60 Hz, single-phase, 15 A; 200–240 VAC, 50 Hz, single-phase, 10 A				115–240 VACm, 50/60 Hz, single-phase, 1200 W	100 VAC, 50/60 Hz, single phase, 15 Amps 115 VAC, 50/60 Hz, single phase, 15 Amps 240 VAC, 50/60 Hz, single phase, 8 Amps

Table 3.3: (*Continued*) Specifications of *ProJet*™ 3500 SD & HD series and *ProJet*™ 5000 and 5500X.

Models	*ProJet*™ 3510 SD	*ProJet*™ 3510 HD	*ProJet*™ 3510 HD*Plus*	*ProJet*™ 3500 HD*Max*	*ProJet*™ 5000	*ProJet*™ 5500X
Dimensions (W × D × H) 3D Printer Crated 3D Printer Uncrated	32.5 × 56.25 × 68.5 in. (826 × 1429 × 1740 mm) 29.5 × 47 × 59.5 in. (749 × 1194 × 1511 mm)	32.5 × 56.25 × 68.5 in. (826 × 1429 × 1740 mm) 29.5 × 47 × 59.5 in. (749 × 1194 × 1511 mm)	32.5 × 56.25 × 68.5 in. (826 × 1429 × 1740 mm) 29.5 × 47 × 59.5 in. (749 × 1194 × 1511 mm)	32.5 × 56.25 × 68.5 in. (826 × 1429 × 1740 mm) 29.5 × 47 × 59.5 in. (749 × 1194 × 1511 mm)	72 × 45.5 × 78 in. (1828 × 1155 × 1981 mm) 60.3 × 35.7 × 57.1 in. (1531 × 908 × 1450 mm)	80 × 48 × 78 in. (2032 × 1219 × 1981 mm) 67 × 35.4 × 65 in. (1700 × 900 × 1650 mm)
Weight 3D Printer Crated 3D Printer Uncrated	955 lbs, 434 kg 711 lbs, 323 kg	955 lbs, 434 kg 711 lbs, 323 kg	955 lbs, 434 kg 711 lbs, 323 kg	955 lbs, 434 kg 711 lbs, 323 kg	1555 lbs, 708 kg 1180 lbs, 538 kg	2550 lbs, 1157 kg 2060 lbs, 934 kg
ProJet™ Accelerator Software	Easy build job setup, submission and job queue management; Automatic part placement and build optimisation tools; Part stacking and nesting capability; Extensive part editing tools; Automatic support generation; Job statistics reporting tools					
Print3D App	Remote monitoring and control from tablet, computers and smartphones				–	–
Network Compatibility	Network ready with 10/100 Ethernet interface					
Client Hardware Recommendation	1.8 GHz with 1 GB RAM (OpenGL support 64 mb video RAM) or higher					1.7 GHz or better with 4GB RAM OpenGL 1.1 Compatible 1280 × 1024 resolution or better
Client Operating System	Windows XP Professional, Windows Vista, Windows 7					Windows 7, Windows 8 or Windows 8.1

Table 3.3: (*Continued*) Specifications of *ProJet*™ 3500 SD & HD series and *ProJet*™ 5000 and 5500X.

Models	ProJet™ 3510 SD	ProJet™ 3510 HD	ProJet™ 3510 HD*Plus*	ProJet™ 3500 HD*Max*	ProJet™ 5000	ProJet™ 5500X
Input Data File Formats Supported	STL and SLC	STL and SLC	STL and SLC	STL and SLC	STL and SLC	STL and SLC
Operating Temperature Range	64–82°F (18–28°C)	64–82°F (18–28°C)	64–82°F (18–28°C)	64–82°F (18–28°C)	64–82°F (18–28°C)	64–82°F (18–28°C)
Noise	<65 dBa estimated (at medium fan setting)					

the preloaded STL files and auto-fix any errors where necessary. The software also helps users to position the parts with its automatic part placement features so as to optimise building space and time. The software also has automatic support generation capabilities to ease the operation of the print of the 3D model. After all details have been finalised, the data is placed in a queue, ready for the *ProJet*™ machine to build the model. The InVision™ Series uses InVision™ Print Client Software instead, and it has similar functions to the *ProJet*™ Accelerator Software for the *ProJet*™ Series.

During the build process, the head is positioned above the platform. The head begins building the first layer by depositing materials as it moves in X-direction. As the machine's printhead contains hundreds of jets, it is able to deposit material fast and efficiently. After a single layer pass is completed, the platform is lowered for the head to work on the next layer while the UV lamp floods the work space to cure the layer. The process is repeated until the part is finished. The schematic of the process is shown in Fig. 3.8.

3.3.4 *Principle*

The principle underlying MJP is the layering principle, used in most other AM systems. MJP builds models using a technique akin to inkjet or phase-change printing, applied in three dimensions. The printhead jets

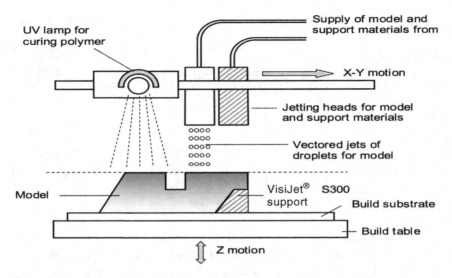

Figure 3.8: Schematic of the MultiJet Modelling System.

are oriented in a linear array builds models in successive layers, each jet applying a special thermo-polymer material only where required. The MJP head shuttles back and forth like an inkjet printer (X-axis), building a single layer of what will soon be a 3D concept model. If the part is wider that the printhead, the platform will then reposition (Y-axis) itself to continue building the layer. The UV lamp flashes with each pass to cure the thermo-polymer deposited. When the layer is completed, the platform is distanced from the head (Z-axis) and the head begins building the next layer. This process is repeated until the entire concept model is completed. The main factors that influence the performance and functionalities of the MJP are the thermo-polymer materials, the UV flood lamp curing, the MJP head (number of jets and their arrangements), the X-Y controls and the Z-controls.

3.3.5 *Strengths and weaknesses*

The strengths of the MJP technology are as follows:

(1) *Efficient and ease of use.* MJP technology is an efficient and economical way to create concept models. The large number of jets

allows fast and continuous material deposition for maximum efficiency. MJP builds models directly from any STL file created with any 3D solid modelling CAD programs and no file preparation is required.

(2) *High precision.* The new MJP 3D Printer provides best-in-class part quality and accuracy with the choice of both high definition and ultra-high definition build modes in a single system. Its resolution of 750 × 750 × 1600 DPI in xyz orientation is one of the highest in the AM industry.

(3) *Cost-effective.* MJP uses inexpensive thermo-polymer material that provides for cost-effective modelling.

(4) *Fast build time.* As a natural consequence of MJP's raster-based design, geometry of the model being built has little effect on building time. Model work volume (envelope) is the singular determining factor for part build time.

(5) *Office friendly process.* As the system is clean, simple and efficient, it does not require special facilities, thereby enabling it to be used directly in an office environment. Due to its networking capabilities, several design workstations can be connected to the machine just like any other computer output peripherals.

The weaknesses of the MJP technology are as follows:

(1) *Small build volume.* The machine has a relatively small build volume as compared to most other high-end AM systems (e.g., *ProJet*® 6000 SD), thus only small prototypes can be fabricated. The *ProJet*™ 3500 HD*Max*, being aimed at jewellery and small components, has the smallest build volume.

(2) *Limited materials.* Materials selections are restricted to 3D systems' *VisiJet*® thermo-polymers. This limited range of materials means that many functionally based concepts that are dependent on material characteristics cannot be effectively tested with the prototypes.

3.3.6 *Applications*

The wide range of uses of the *ProJet*™ machines, include applications for concept development, design validation, form and fit analysis, production of moulding and casting patterns, direct investment casting of jewellery and other fine feature applications. Specifically for the *ProJet*™ machines, they can be used in the dental laboratory.

3.3.7 *Research and development*

In 2013, 3D Systems acquired a portion of Xerox's R&D group which deals with product design, engineering and chemistry. The acquisition brought in an additional 100 engineers and contractors from Xerox into 3D Systems' R&D department. Xerox has long served as a key partner in 3D Systems' most popular line of inkjet-based 3D printers, the *ProJet* series. With this acquisition, the MJP technology will evolve and it serves as a good indication that the *ProJet* series will get bigger, better and cheaper.

3.4 EnvisionTEC's *Perfactory*®

3.4.1 *Company*

The company was founded as Envision Technologies GmbH in August 1999 and was re-organised as EnvisionTEC GmbH in 2002 by Ali El Siblani. EnvisionTEC provides 3D printing systems and solutions to serve customers in a wide variety of applications. EnvisionTEC also deals with software and materials development to increase productivity and cost-effectiveness. The company has dual headquarters in Gladbeck, Germany and Dearborn, Michigan, USA.

3.4.2 *Product*

Perfactory® (see Fig. 3.9) is the AM system built by EnvisionTEC and it is a versatile system suitable for an office environment.[16] *Perfactory*®

Figure 3.9: *Perfactory*® 3 Mini Multi.
Courtesy of EnvisionTEC

Table 3.4(a): *Perfactory*® 3 Mini Multi Lens specifications.

System	*Perfactory*® 3 Mini Multi Lens		
Lens system, focal length, mm (in.)	60 mm (2.36")	75 mm (2.95")	85 mm (3.35")
Build envelope XYZ, mm, (in.)	84 × 63 × 230 mm 3.31" × 2.48" × 9.06"	59 × 44 × 230 mm 2.32" × 1.73" × 9.06"	44 × 33 × 230 mm 1.73" × 1.3" × 9.06"
Dynamic Voxel thickness Z, μm (in.)	15–150 μm (0.0006–0.006)		
Voxel size XY W/ERM, μm (in.)	30 μm (0.0012")	21 μm (0.0008")	16 μm (0.0006")

Courtesy of EnvisionTEC GmbH
+Double *Perfactory*® Resolution

Table 3.4(b): *Perfactory*® 4 Standard Series System specifications.

System	Perfactory® 4 Series	
	Perfactory® 4 Standard	**Perfactory® 4 Standard XL**
Build envelope XYZ (mm)	160 × 100 × 230 mm 6.3" × 3.9" × 6.3"	192 × 120 × 230 7.6" × 4.7" × 6.3"
Native Voxel Size XY, μm	83 μm 0.0033"	100 μm 0.0039"
ERM Voxel size XY, μm	42 μm 0.0017"	50 μm 0.0020"
Dynamic Voxel thickness Z, μm	25–150 μm 0.0010"–0.0059"	25–150 μm 0.0010"–0.0059"
Projector Resolution	1920 × 1200 pixels	1920 × 1200 pixels

Courtesy of EnvisionTEC GmbH

Table 3.4(c): *Perfactory*® SXGA⁺ Standard and Standard UV systems specifications.

System	*Perfactory*® 4 Mini with 60 mm lens	*Perfactory*® 4 Mini with 75 mm lens	*Perfactory*® 4 Mini XL with 60 mm lens	*Perfactory*® 4 Mini XL with 75 mm lens
Build envelope XYZ (mm)	64 × 40 × 230 mm 2.5" × 1.6" × 9.1"	38 × 24 × 230 mm 1.5" × 0.9" × 91"	115 × 72 × 230 mm 4.5" × 2.8" × 9.1"	84 × 52.5 × 230 mm 3.3" × 2.1" × 9.1"
Native Voxel size XY, μm	33 μm 0.0013"	19 μm 0.0007"	60 μm 0.0023"	44 μm 0.0017"
ERM Voxel size XY, μm	17 μm 0.0007"	10 μm 0.0004"	30 μm 0.0012"	22 μm 0.0009"
Dynamic Voxel thickness Z, μm	15–150 μm 0.0006"–0.0059"	15–150 μm 0.0006"–0.0059"	15–150 μm 0.0006"–0.0059"	15–150 μm 0.0006"–0.0059"
Projector Resolution	1920 × 1200 pixels	1920 × 1200 pixels	1920 × 1200 pixels	1920 × 1200 pixels

Courtesy of EnvisionTEC GmbH

requires a file in an STL file format and the STL data are transferred to the system to build the model using the included software package. Resins are cured by photopolymerisation, but *Perfactory*® uses a different approach from traditional stereolithography in curing the resins. The photopolymerisation process is created by an image projection technology called Digital Light Processing Technology (*DLP*™) from

Texas Instruments and it requires mask projection to cure the resin. The standard system alone can achieve resolutions between 25 and 150 μm, material dependent. Additional components or devices such as the Mini Multi Lens system and the Enhanced Resolution Module (ERM) enable designers and manufacturers to build smaller figures, which require high surface quality. The *Perfactory*® 3 Mini Multi Lens, fitted with ERM and an 85-mm lens is able to create high quality parts with a voxel size as low as 16 μm. The specifications of *Perfactory*® systems are shown in Tables 3.4(a)–(c).

3.4.3 Process

To build the part in a systematic manner, *Perfactory*® undergoes a simple process (see Fig. 3.10) where the 3D model of a solid model is first created with a commercial CAD system. For medical applications, data acquired with MRT or CT systems can be processed directly. The 3D data model in the STL format acquired is voxelised within the *Perfactory*® software and each voxel data set is converted into a bitmap file with which the mask image is generated. The bitmap image consists of black and white where white represents the material and black represents the void as well as grey which represents partial curing of the voxel. When the image is projected onto the resin with *DLP*™, the illuminated white portion will cure the resin while the black areas will not and grey area on inner and outer contours will be partially cured in depth thereby eliminating the layering effect.

With the embedded operating system in *Perfactory*®, it can operate independently and is monitored by the device driver software installed on

Figure 3.10: *Perfactory*® process. (a) 3D data model (STL), (b) Voxel data model, (c) Mask projection, (d) 3D solid model.
Courtesy of EnvisionTEC GmbH

the PC. The software provides two types of modes, Auto and Expert. The Auto mode allows direct conversion of 3D-CAD data to STL format and other required setups are programmed automatically. The Expert mode is specially programmed for advanced users to offer them with choice of manual setup according to their experience, needs and preferences.

Unlike almost all other AM systems that build the model from bottom up, *Perfactory*® builds the model top down (see Fig. 3.11). The build plate or carrier is first immersed into a shallow trough (basement) of acrylate-based photopolymer resin sitting on a transparent contact window. The mask is projected from below the build area onto the resin to cure it. Once the resin is cured, the build platform is raised a single white voxel depth, the voxel thickness being dependent on the material. While the platform is raised, it peels the model away from the transparent contact window, thus allowing fresh resin to flow in through capillary action. The next exposure is then applied and the part is then built in a similar manner. The whole cycle can take as little as 15 seconds without the need for planarisation or levelling for each voxel set cured.

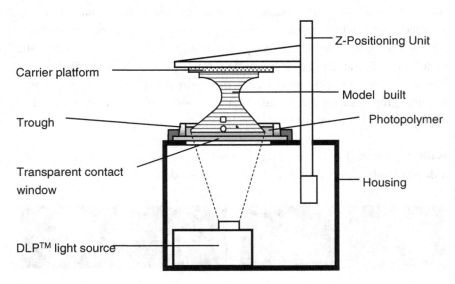

Figure 3.11: Schematic of the *Perfactory*® build process with *DLP*™ technology.
Courtesy of EnvisionTEC GmbH

The two key differences of the *Perfactory®* with other AM systems are the use of mask projection for photopolymerisation and the part is moved upwards with each completed cured layer instead of downwards in other systems. Once the model is built, the user simply has to peel the model off from the platform as the model is stuck to the carrier platform during the entire process.

3.4.4 *Principle*

Perfactory® uses the basic principles of stereolithography by undergoing the following stages:

(1) Parts are built from acrylate photopolymer and the user is able to select different material properties with different material colours. Resins are cured when exposed to a mask projection image using *DLP*™ technology from Texas Instruments.
(2) Every completion of cured layer is moved away from the build trough containing the resin vertically upwards by a precision linear drive. This is due to the projection system being integrated at the bottom of the AM system. Also, the fabricated part requires minimal support and removing the model from the transparent platform is easy.

3.4.4.1 *DLP technology*

DLP™ is a projection technology invented by Dr. Larry Hornbeck and Dr. William E. "Ed" Nelson of Texas Instruments in 1987. In *DLP*™ technology, the key device is the Digital Micromirror Device (DMD), the producer of image. DMD is an optical semi-conductor and each DMD chip has hundreds of thousands of mirrors arranged in a rectangular array on its surface to steer the photons with great accuracy. This means that each mirror is represented as one pixel in a projected image and therefore, the resolution of an image depends on the number of mirrors.

The mirrors in the DMD are made of aluminium and are 16 µm in size. Each individual mirror is connected to two support posts where it can be rotated ±10–12° in an ON or OFF state. In the ON state, the light source is reflected from the mirrors into the lens making pixels on the screen. For the OFF state, the reflected light is redirected to the other direction allowing the pixels to appear in a dark tone. During each mask projection, the cross section of each layer is projected by mirrors in the ON state and resins are cured by the visible light projected from below the transparent contact window.

Every single mirror is mounted on a yoke by compliant torsion hinges with its axle fixed on both ends and able to twist in the middle. Due to its extremely small scale, damages hardly occur and the DMD structure is able to absorb shock and vibration thus providing high stability.

Position control of each mirror is done by two pairs of electrodes as positioning significantly affects the overall image of the cross-section layer. One pair is connected to the yoke while the other is connected to the mirror and every pair has an electrode on each side of the hinge. Most of the time, equal bias charges are applied to both sides to hold the mirror firmly in its current position. In order to move the mirrors, the required state has to load into a static random access memory (SRAM) cell connecting to the electrodes and mirror. Once all the SRAM cells have been loaded, the bias voltage is removed, allowing every individual movement of the mirrors through released charges from the SRAM cell. When the bias is restored, all of the mirrors will be held in their current position to wait for the next loading into the cell. Bias voltage enables instant removal from the DMD chip so that every mirror can be moved together providing more accurate timing while requiring a lower amount of voltage for the addressing of the pixel.

3.4.5 *Strengths and weaknesses*

The main strengths of *Perfactory*® systems are:

(1) *High building speed.* The use of a mask image directly exposed to the resin enables a part to be built at speeds as quick as

approximately 15 mm height per hour at 100 μm pixel height. This speed is independent of part size and geometry and is one of the fastest systems in the AM market.

(2) *Office friendly process. Perfactory*® allows operation in an office environment as its foot print is under 0.3 m². The curing of the photopolymer does not use UV light and there is no need for special facility. It operates with low noise and zero dust emissions.

(3) *Small quantity of resin during build.* The shallow trough of resin means that the amount of material in use at any one time is small (about 200 mL). This means that should a number of different resins be required, they can be swapped out easily with minimal wastage.

(4) *No wiper or leveller.* When the carrier platform is raised with the model, there is no need for planarisation or levelling. This eliminates the possibility of causing problems to the stability of the parts during the build, for example, a wiper damaging a small detail on the part during the wiping action.

(5) *Less shrinkage.* Due to immediate curing of a controlled layer (based on the voxel thickness) there is less shrinkage during the process.

(6) *Additional components. Perfactory*® is able to build even higher quality and tougher parts with the use of additional components which can be integrated into the system.

The main weaknesses of *Perfactory*® systems are:

(1) *Limited building volumes.* Structures are built from the bottom of the build chamber and stuck to the carrier platform; this limits the size of the build.

(2) *Peeling of completed part.* The user has to peel the completed model from the build platform as the model is built on a movable carrier platform which moves vertically upwards. Care has to be taken so as not to damage the model during the peeling process.

(3) *Requires post-processing.* After the model is completed, cleaning and post-processing, sometimes including post-curing are required.

3.4.6 *Applications*

The application areas of the *Perfactory*® systems include the following:

(1) Concept design models for design verification, visualisation, marketing and commercial presentation purposes.
(2) Working models for assembly purpose, simple functional tests and for conducting experiments.
(3) Master models and patterns for simple moulding and investment casting purposes.
(4) Building and limited production of completely finished parts.
(5) Medical[17] and dental applications. Creating exact physical models of patient's anatomy from CT and MRI scans.
(6) Jewellery applications. Creating designs for direct casting or moulding.

3.4.7 *Research and development*

EnvisionTEC has been researching on building larger build envelope machines and new casting materials. Recently the company announced two new casting materials called Epic and Epic M. Epic is a wax-based material and offers the quality and detail of the plastic casting materials but with the properties and adaptability of its wax products.[18]

3.5   RegenHU's 3D Bioprinting

3.5.1 *Company*

RegenHU Ltd was founded in November 2011. The company specialises in making bioprinters and producing 3D organomimetic models for tissue engineering. RegenHU Ltd is headquartered in Z.i. du Vivier 22, CH-1690 Villaz-St-Pierre, Switzerland.

3.5.2 *Products*

RegenHU Ltd produces a wide range of 3D printers to produce biotissues and biomaterials. There are two main models which are the

3DDiscovery® and *BioFactory®*. *3DDiscovery®* printer (see Fig. 3.12) is designed to further study the potential of 3D tissue engineering through the bioprinting approach. Its innovation lies in the spatial controlling of cells and morphogens in a 3D scaffold. The system helps in constructing organotypic *in vitro* models of soft and hard tissues. Meanwhile, *BioFactory®* (see Fig. 3.13) is designed for scientists and specialists to pattern cells, biomolecules and a range of soft and rigid materials in desirable 3D composite structures. Both *3DDiscovery®* and *BioFactory®* use *BioInk™* as the building materials because *BioInk™* is a semi-synthetic hydrogel supporting cell growth of different cell types. Specifications of *3DDiscovery®* printer are summarised in Table 3.5.

Figure 3.12: *3DDiscovery®* printer.   Figure 3.13: *BioFactory®* printer.
Courtesy of RegenHU Ltd   Courtesy of RegenHU Ltd

Table 3.5: Specifications of *3DDiscovery®* and *BioFactory®* printer.

Model	*3DDiscovery®* printer	*BioFactory®* printer
Working range	130 × 90 × 60 mm	60 × 60 × 60 mm
Precision	± 5 μm	
Modular printhead concept	General	
Temperature control	From 0°C up to 80°C (substrate holder, medias)	From 5°C up to 80°C (substrate holder, medias)
Overall dimensions (W × L × H)	600 × 700 × 670 mm	1370 × 1030 × 2400 mm

Nano liter dispensing resolution, minimal dead volume; printing under physiological conditions.

3.5.3 *Process*

RegenHU bioprinting applies the following steps to creating the tissue or cell:

(1) The CAD file of building process is formed in G-Code format. STL importation of 3D objects are generated through medical imaging (computer tomography, MRI) and/or CAD/CAM systems.
(2) RegenHU's bioprinters produce the tissue and cells layer upon layer.
(3) Natural tissue is assembled by the 3D printed cell and extracellular matrix.
(4) The final product is removed from the working plate.

3.5.4 *Principle*

The design of the RegenHU's 3D bioprinters is created for the purpose of tissue engineering applications. Tissues are built up layer upon layer. Cells, signal molecules and biomaterials create a highly dynamic network of proteins and signal transduction pathways. The printed cell closely mimics the natural tissue. Its well-defined structure enables the control and study of biological and mechanical cell/molecule interaction processes.

3.5.5 *Materials*

Generally, the materials used are dependent on the cell or tissues being printed. Cells, proteins, bioactives, thermo-polymers, natural hydrogels such as collagen, hyaluronic acid, chitosan and gelatine can be used in the RegenHU bioprinters. RegenHU provides two special materials dedicated for their bioprinters:

3.5.5.1 *BioInk®*

BioInk® (see Fig. 3.14) is a semi-synthetic hydrogel that promotes cell growth of different cells by providing cell adhesion sites and mimicking

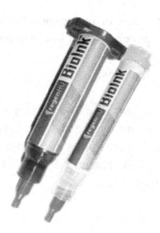

Figure 3.14: The *BioInk®*.
Courtesy of RegenHU Ltd.

the natural extracellular matrix. *BioInk®* is available as a ready-to-use solution to print 3D tissues models.

3.5.5.2 *OsteoInk®*

OsteoInk® is a ready-to-use calcium phosphate paste for structural engineering dedicated to *BioFactory®* and *3DDiscovery®* tissue printers. OsteoInk's chemical composition such as Hydroxyapatite or tricalcium phosphate is optimal for hard tissue engineering such than bone and cartilage. 3D printing enables the freeform fabrication with controlled pore structure. *BioFactory* and *3DDiscovery* enable composite tissue printing by loading or filling the pores with bioactives such as cells, proteins, blood derivate or hydrogels.

3.5.6 *Strengths and weaknesses*

RegenHU's bioprinters have the following strengths:

(1) *Unique technology*: Researchers and scientists can design and produce pattern cells, biomolecules and a range of soft and rigid materials in desirable composite structures.

(2) *Cost-effective*: The cost of using such bioprinters is relatively low, which allows researchers to do a few attempts.

It has the following weaknesses:

(1) *Small building volume*: The bio printer's working range is only 130 × 90 × 60 mm, limiting the size of the build.
(2) *Sterile environment*: Keeping the bioprinting process under a sterile environment is crucial as materials include biomaterials, biochemical and living cells.

3.5.7 *Applications*

The general applications areas are given as follows:

(1) *Skin model (soft tissue engineering)*: RegenHU's *BioFactory*® can produce full thickness skin models consisting of dermal and epidermal layers. This helps researchers to further study on them.
(2) *Drug discovery*: RegenHU's *BioFactory*® helps to provide an alternative for researchers to study or analyse the induced effects of active substances on highly dynamic networks of proteins and signal transduction pathways in tissues, cell-cell and cell-extracellular matrix interactions. Traditionally, this kind of research can only be done in a 2D manner.
(3) *System engineering*: With RegenHU's bioprinters, the risks associated with the reproducibility, traceability and quality control of drugs developments are mitigated.

3.5.8 *Research and development*

RegenHU is focusing on the development of the hardware of its bioprinters. The materials used by the machines are mainly developed by their key consumers and R&D institutes worldwide.

3.6 Rapid Freeze Prototyping

3.6.1 *Introduction*

Most of the existing AM processes are relatively expensive and many of them generate substances such as smoke, dust, hazardous chemicals and so forth, which are harmful to human health and the environment. Continuing innovation is essential in order to create new rapid prototyping processes that are fast, clean and of low cost.

Dr. Ming Leu of the University of Missouri-Rolla, has been developing a novel, environmentally benign AM process that uses cheap and clean materials and can achieve good layer binding strength, fine build resolution and fast build speed. They have invented such a process called rapid freeze prototyping (RFP)[19]; it makes 3D ice parts layer by layer by freezing water droplets.

3.6.2 *Objectives*

The fundamental study of this process has three objectives:

(1) Developing a good understanding of the physics of this process, including the heat transfer and flow behaviour of the deposited material in forming an ice part.
(2) Developing a part building strategy to minimise part build time, while maintaining the quality and stability of the build process.
(3) Investigating the possibility of investment casting application using the ice parts generated by this process.

3.6.3 *Process*

The experimental AM system[20] in Fig. 3.15 consists of the following mechanical mechanism: (1) a 3D positioning subsystem; (2) a material depositing subsystem (see Fig. 3.16); (3) a freezing chamber and (4) electronic control device. Tedious modelling and analysis efforts[21] have been carried out to understand and improve the behaviour of material solidification and rate of deposition during the ice part building process. The process starts with software which receives STL file from CAD software and generates the sliced layers of contour information under a CLI file. Together with a CLI file, process parameters like nozzle transverse speed, environment temperature and fluid viscosity are then further processed to generate the NC code. The NC code is sent to the experimental system to control the fabrication of the ice parts.

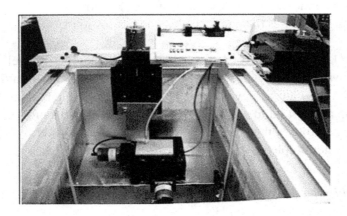

Figure 3.15: Experimental system of RFP.

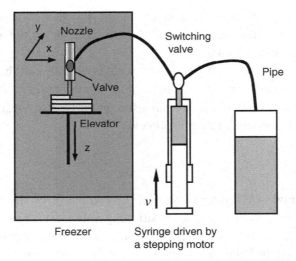

Figure 3.16: Building environment and water extrusion subsystem.

Figure 3.17: An ice part before removal of support material and after removal.

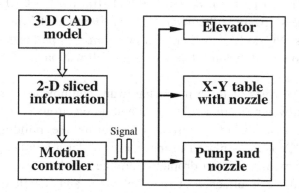

Figure 3.18: Control system schematics.

The 3D positioning system consists of an XY table and a Z-elevator and it is seated in the freezer at a low temperature of −20°C or lower. The depositing nozzle is mounted on the Z-elevator whereby water droplets are extracted from the nozzle onto the substrate. Once the first layer solidifies, the nozzle will eventually level up and continue depositing for the next layer.

During the fabrication process, building and support solutions are used. The building solution will be water and the support solution is the eutectic sugar solution ($C_6H_{12}O_6$–H_2O) of a melting temperature −5.6°C.[22] Due to the difference in each solution's melting temperature, the built part is then placed in an environment between 0°C and the melting point of the support material to melt the support material (see Fig. 3.17). Experiments have also been conducted where small parts are fabricated with radius of 0.205 mm and thin wall of 0.48 mm in thickness.[23] Fig. 3.18 shows the control system schematics for the system.

3.6.4 *Principles*

The rapid freezing prototyping system is based on the principle of using a water freezing method to build an icy cold prototype. Besides being environmentally friendly, the water solution can also be easily recycled. Before the frozen part can be built, a STL file from a CAD design is required. The software will convert the STL file to a CLI file and finally to NC codes. NC codes together with other process parameters are then transferred to a motion control card to begin fabrication.

The system consists of three major hardware equipments: a positioning subsystem, a water ejecting subsystem and an electronic control device. Each of these holds an important role during the building process. Firstly, the XY table is placed in the freezer under a low temperature in order for the extracted water droplets from the nozzle to freeze in a short time. The nozzle is attached at the Z-elevator and it will move in an upwards direction after every completed layer. For the first layer of the ice part, only water droplets are deposited. Every water droplet does not solidify immediately upon deposition as the water droplets are spread and form a continuous water line. The water line will then be frozen by convection through the cold environment and conduction from the previous frozen layer rapidly.

During each experimental process for building an ice part, process parameters set in the water ejection subsystem and electronic control device are important as they affect the overall fabrication accuracy. The key parameters include the ambient build temperature, scan speed and water feed rate. The ratio of material flow to XY movement speed is kept at the best value to prevent discontinuity in the freezing strands. Each layer thickness and smoothness is determined by the adjustment of nozzle scanning speed and water feed rate rather than the mechanical mechanism.[24]

3.6.5 *Strengths and weaknesses*

RFP has the following strengths:

(1) *Low running cost.* The RFP process is cheaper and cleaner than all the other AM processes. The energy utilisation of RFP is low compared to other AM processes such as laser stereolithography or SLS.

(2) *Good accuracy.* RFP can build accurate ice parts with excellent surface finish. It is easy to remove the RFP made ice part in a mould making process by simply heating the mould to melt the ice part.

(3) *Good building speed.* The build speed of RFP can be significantly faster than other AM processes, because a part can be built by first depositing water droplets to form a waterline and then filling in the enclosed interior with a water stream (see Fig. 3.19). This is possible due to the low viscosity of water. It is easy to build colour and transparent parts with the RFP process.

(4) *Environmentally friendly materials.* Both the building material and support material are clean and non-toxic when handled.

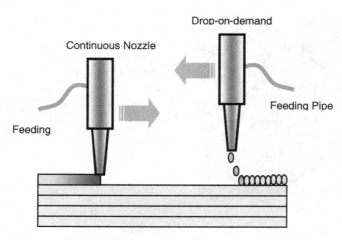

Figure 3.19: Two material extrusion methods.

(5) *Suitability in investment casting.* Eliminating problems created from traditional wax patterns in investment casing like pattern expansion and shelling cracking during pattern removal.

On the other hand, RFP has also the following weaknesses:

(1) *Requires cold environment.* The prototype of RFP is made of ice and hence it cannot maintain its original shape and form at room temperature.
(2) *Needs additional processing.* The prototype made with RFP cannot be used directly but has to be subsequently cast into a mould and so on and this increases the production cost and time.
(3) *Repeatability.* Due to the nature of water, the part built in one run may differ from the next one. The composition of water is also hard to control and determine unless tests are carried out.
(4) *Post-processing.* The final part is required to be placed in a lower temperature environment to remove the support material.

3.6.6 *Potential applications*

(1) *Part visualisation.* Parts can be built for the purpose of visualisation. Examples can be seen in Fig. 3.20.

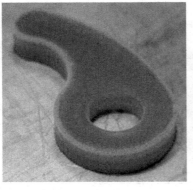 

(a)   (b)

Figure 3.20: Examples of visualisation parts: (a) Solid link rod (10 mm height) and (b) contour of a link rod.

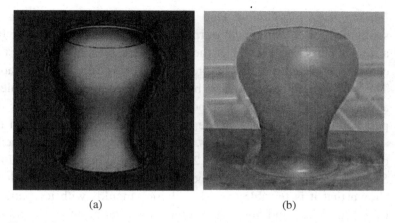

(a)   (b)

Figure 3.21: Ice sculpture: (a) the CAD model and (b) the RFP fabricate d part.

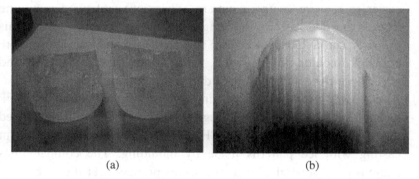

(a)   (b)

Figure 3.22: (a) UV silicone mould made by ice pattern and (b) urethane part made by UV silicone mould.

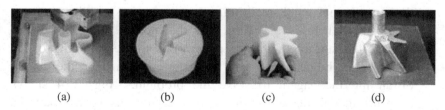

(a)   (b)   (c)   (d)

Figure 3.23: (a) Making the ice part by RFP, (b) UV silicone mould made by ice part, (c) urethane part made by UV silicone mould and (d) metal part made by UV silicone mould.

(2) *Ice sculpture fabrication.* One application of RFP is making ice sculptures for entertainment purposes. Imagine how much more fun a dinner party can provide if there are colourful ice sculptures that can be prescribed one day before and they vary from table to table. This is not an unlikely scenario because RFP can potentially achieve a very fast build speed by depositing water droplets only for the part boundary and filling in the interior with a water stream in the ice sculpture fabrication process. Fig. 3.21(a) shows the CAD model of an ice sculpture, while Fig. 3.21(b) shows the part made by RFP.

(3) *Silicone moulding.* The experiments on UV silicone moulding have shown that it is feasible to make silicone moulds with ice patterns (see Fig. 3.22) and further make metal parts from the resulting silicone moulds (see Fig. 3.23). The key advantage of using ice patterns instead of plastic or wax patterns is that ice patterns are easier to remove (without pattern expansion) and no demoulding step is needed before injecting urethane or plastic parts. This property can avoid demoulding accuracy loss and can allow more complex moulds to be made without the time-consuming and experience-dependent design of demoulding lines.

(4) *Investment casting.* A promising industrial application of RFP technology is investment casting. DURAMAX recently developed the freeze cast process (FCP), a technology of investment casting with ice patterns made by moulding. The company has demonstrated several advantages of this process over the competing wax investment and other casting processes, including low cost (35%–65% reduction), high quality, fine surface finish, no shell cracking, easy process operation and faster run cycles. Additionally, there is no smoke and smell in investment casting with ice patterns. Apparently, there is a finding that the alcohol base binder can damage the surface of the ice pattern due to the solubility of ice and alcohol binder. Research has been ongoing for the search of an interface agent to solve the problem.[25] Fig. 3.24 shows a metal model made by investment casting using ice patterns from RFP.

Figure 3.24: Metal model made by investment casting with ice pattern.

3.7 Optomec's Aerosol Jet Systems

3.7.1 *Company*

Optomec is a privately held company founded in 1982, with corporate headquarters located in Albuquerque, New Mexico, and with an Advanced Applications Lab and New Product Development Centre located in Saint Paul, Minnesota. Since 1997, Optomec has focused on commercialising LENS® AM technology for 3D Printed Metals, which was originally developed by Sandia National Laboratories. Optomec delivered its first commercial LENS® system to Ohio State University in 1998. In 2004, Optomec released their first commercial Aerosol Jet® system for printed electronics. Optomec Inc. is located at 3911 Singer Boulevard NE, Albuquerque, NM 87109, USA.

3.7.2 *Aerosol Jet systems*

Optomec offers three models of standard Aerosol Jet systems for printed electronics each targeted to meet specific market needs. The Aerosol Jet 200 Series is a professional grade benchtop system ideally suited for universities, ink developers and others exploring the benefits of AM technology for electronic applications. The Aerosol Jet 300 System Series, with its higher print accuracy motion system and larger work area, is ideally suited for developing next generation printed electronics and biologics processes/prototypes as well as low-volume production. For production level printed electronics, Optomec offers the Aerosol Jet 5X system for applications such as complex moulded interconnect devices, embedded sensors, smartphone antennas and low-volume production. An Aerosol Jet Print Engine is also available for integration with industrial automation platforms for high-volume production applications. The Aerosol Jet Print Engine and standard systems all utilise the same industry proven Aerosol Jet core deposition technology which enables processing of wide variety of electronic materials and printing of a wide range of feature sizes on both 2D and 3D substrates. Table 3.6 compares the standard features available in each Aerosol Jet System model.

Table 3.6: Aerosol Jet Systems features comparison.

Standard Features	Aerosol Jet System Models		
	AJ 200	AJ 300	AJ 5X
Process	Aerosol	Aerosol	Aerosol
Build volume (mm)	200 × 200	300 × 300	200 × 300 × 200
Atomisers	Ultrasonic, with Pneumatic option	Ultrasonic and Pneumatic	Ultrasonic and Pneumatic
Fine feature print head	With shutter	With heater and shutter	With heater and shutter
Wide feature print head	With shutter (optional)	With shutter (optional)	With shutter
Axis motion	*X, Y*	*X, Y*	*X, Y, Z*
Motion accuracy	± 25 microns	± 6 microns	± 10 microns
Safety enclosure	No	Yes	Yes
System and process control software	Yes	Yes	Yes
Tool path generation software	Yes	Yes	Yes
Stainless steel cabinet with granite table top	No	Yes	Yes
Materials	See Table 3.7 for Materials Supported by the AJ Process		

3.7.3 *Aerosol Jet process*

Aerosol Jet Systems atomise the source material into a dense aerosol mist composed of droplets with diameters between approximately 1 and

5 microns. The resulting aerosol stream contains particles that are further treated on the fly to provide optimum process flexibility. The material stream is then focused using a flow guidance deposition head, which creates an annular flow of sheath gas to collimate the aerosol. The co-axial flow serves to focus the material stream. The resulting high-velocity material stream is deposited onto planar and 3D substrates, creating features ranging from 10 microns to centimetres in size as shown in Fig. 3.25. A motion control system allows the creation of complex patterns on the substrate. For low-temperature substrates, deposited material can be laser sintered to achieve properties near those of the bulk material without damage to the surrounding substrate. The Aerosol Jet process can print a wide variety of materials on a wide variety of substrates. Typically, if a source material can be suspended in a solution and atomised, it can be printed with the Aerosol Jet process. Table 3.7 provides a list of inks that have been printed on various substrates using Aerosol Jet technology.

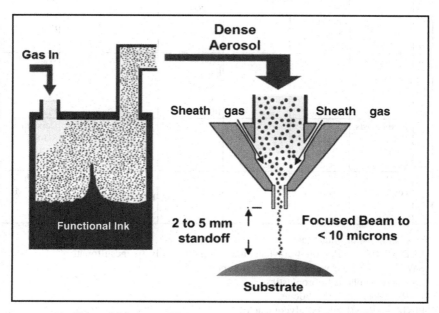

Figure 3.25: Aerosol Jet process with ultrasonic atomiser.

Table 3.7: Supported materials using the Aerosol Jet process.

Aerosol Jet Materials Matrix		Substrates					
		Polyester	Polyimide	Glass	C-Si	ITO	Metals
Inks	Conductor	Silver, Gold, PEDOT: P-SS,[b] SWCNT[c]	Silver, Gold, Copper, SWCNT[c]	Silver, Copper, PEDOT: P-SS,[b] SWCNT[c]	Fritted Silver	Silver, Gold, Aluminium, PEDOT: PSS [b]	Silver, Gold, Aluminium, PEDOT: PSS[b]
	Resistor	Asahi FTU series[a]	Asahi TU series,[a] SWCNT,[c] M-2031-pol[a] (Polyimide)	Asahi TU series,[a] SWCNT,[c] M-2031-pol[a] (Polyimide)			Asahi TU series[a]
Inks	Dielectric Adhesive	Loctite 3492,[a] Norland 65,[a] Gersteltec SU-8 GM 1010,[a] Polyimide, Teflon[a] AF,[d] Luvitec[a] PVP,[e] Ablestik[a]	Loctite 3492,[a] Norland 65,[a] Teflon AF,[d] Luvitec[a] PVP[e]	Norland 65,[a] Gersteltec SU-8 GM 1010,[a] Polyimer, Asahi CX-16,[a] Teflon[a] Dispersion, Luvitec[a] PVP[e]		Norland 65,[a] Polyimer, Luvitec[a] PVP[e]	Norland 65,[a] Loctite 3492,[a] Gersteltec SU-8 GM 1010,[a] Teflon[a] AF,[d] Luvitec[a] PVP[e]
	Biological			Water based, solvent based, PLGA[f] 5%			Water based, solvent based, PLGA[f] 5%
	Etch Resists Catalysts, Etch Chemicals				Enlight[a]		Gersteltec SU-8 GM 1010,[a] Acids
	Semi-conductor	P3HT,[g] SWCNT[c]	P3HT,[g] SWCNT[c]	P3HT,[g] SWCNT[c]	Enlight[a]		Gersteltec SU-8 GM 1010,[a] Acids

[a] Product name or brand name.

[b] PEDOT: PSS — Poly(3, 4-ethylenedioxythiophene) polystyrenesulfonate.

[c] SWCNT — Single-walled carbon nanotubes.

[d] AF — Amorphous fluoroplastics.

[e] PVP — Polyvinylpyrrolidone.

[f] PLGA — Poly(lactic-co-glycolic acid).

[g] P3HT — Poly(3-hexylthiophene-2, 5-diyl) regioregular.

3.7.4 *Aerosol Jet's strengths and weaknesses*

The key strengths of Aerosol Jet systems are as follows:

(1) *Planar and non-planar (3D) printing capabilities.* Enables functionalised circuitry on 3D substrates.
(2) *Printed feature sizes ranging from 10 microns to centimetres.* Addresses shrinking electronic package size challenges and enables denser component spacing than previously possible.
(3) *Thin layer deposits from 100 nm.* Targeted at precise coating and multi-material applications, like solid oxide fuel cell (SOFC), using less material than traditional manufacturing methods.
(4) *A wide range of inks and substrates.* From nanomaterials, such as silver and gold, to polymers, epoxies, etchants, dopants and biological materials have all been dispensed with Aerosol Jet onto substrates such as Kapton, polycarbonate, glass, silicon, fabrics and so on.
(5) *Low-temperature processing.* Enable Aerosol Jet dispensing at ambient temperatures without the need for specialised chambers or temperature controlled environments.

The limitations of Aerosol Jet systems are as follows:

(1) Requirement for particle sizes of less than 1 micron will limit use of some current conductive inks.
(2) Current nozzle spacing of approximately 2.5 mm will limit use for certain applications that require higher resolution.

3.7.5 *Aerosol Jet applications*

The Aerosol Jet system allows "high-tech" printed electronics applications such as:

- Fine Feature (10 micron) to Wide Feature Circuits (1 cm)
- Printed passive components (resistors, capacitors)

- Printed active components (antennas, sensors and thin-film transistors)
- Printed batteries and fuel cells
- 3D Printed Electronics (stacked dies interconnects, cell phone antennas and sensors; see Fig. 3.26)
- Printed biologics

Fully printed components and circuits eliminate the need for mounting and soldering physical devices to a carrier board. Using Aerosol Jet technology, components can even be printed directly onto or inside 3D structures, such as the wing of an aircraft or a cell phone case, thereby eliminating the need for separate circuit boards. This capability to print electronics on 3D surfaces opens up new packaging opportunities for lighter, thinner electronic devices with increased functionality.

Figure 3.26: 3D Interconnects (left), 3D Printed antenna (centre), 3D Creep Sensor on Ti Blade (right).
Courtesy of Optomec

3.7.6   *Research and development*

Optomec's R&D focus is on optimising Aerosol Jet technology for high-volume repair and manufacturing application. Aerosol Jet Print Engine technologies are ideally suited for high-volume production. For example, the Aerosol Jet Print Engine can be scaled to enable simultaneous printing with print heads on a single automation platform. Optomec users are currently qualifying Aerosol Jet Print Engines in high-volume production.

3.8 Two-Photon Polymerisation

3.8.1 *Introduction*

Two-photon polymerisation (TPP) is the first AM technology capable of fabricating true 3D structures without undergoing the layer-by-layer process. TPP microfabrication utilises ultrashort laser pulses. When focused into the volume of photosensitive material (or photoresist), the pulses initiate TPP via two-photon absorption and subsequent polymerisation. TPP enables the fabrication of any computer-generated 3D structures by direct laser "recording" into the volume of a photosensitive material.[26]

3.8.2 *Process*

The process of TPP is achieved through free radical polymerization of the liquid resin by a photolytic process. The photons from the light source excite the photoinitiator molecules in the resin, creating free radicals, which initiate the polymerisation process. These free radicals join the monomers in the liquid resin to form chain radicals that grow longer by adding monomers to the chain. The process of polymerisation ends when chain free radicals join to form a molecule with no radical activity. The mechanism by which polymerisation by two photon absorption is outlined below.[26]

Let:
I: indicate the photoinitiator
R: free radical
M: monomer-oligomer
Mn: polymer
R–(Mn): chained radical
hv: photon
•: unpaired electron creating radical molecule

Initiation:

$I + 2hv \rightarrow 2R\bullet$

$R\bullet + M \rightarrow RM\bullet$

Propagation:

$R-(Mn)_x\bullet + M \rightarrow R - (Mn)_{x+1}\bullet$

Chain radical of length $\times$ + monomer $\rightarrow$ chained radical of length $\times$+1

Termination:

$-M\bullet + \bullet M - \rightarrow -M-M-$

The TPP process requires high radical density, thus an fs-pulsed laser is needed to combine with high numerical aperture (NA) focusing to provide a high light intensity spot. Fig. 3.27 shows a typical fabrication setup to generate microstructures. Near infrared (NIR) fs laser pulses generated in a Ti:Sapphire laser are converted to the visible by using an optical parametric oscillator (OPO). A neutral density (ND) filter is introduced into the beam path allowing for adjustments in beam intensity. The telescope arrangement leads to uniform illumination over the back aperture of an oil immersion objective. The beam is focused into a sample to initiate polymerisation. The sample, a photo-sensitive liquid resin, is sandwiched between two coverslips and mounted at a 3D piezoelectric scanning stage. The movement of the scanning stage is pre-programmed to form different micro-structures and controlled by a computer. The fabrication process is monitored by a charge-coupled device camera. After fabrication, the unpolymerised photoresist is washed out with ethanol and the desired 3D microstructure can be achieved.

3.8.3 *Principle*

TPP was experimentally reported for the first time in 1965 by Pao and Rentzepis as the first example of multi-photon excitation-induced photochemical reactions. Photopolymerisation can be defined as a chemical reaction that converts molecules of low molecular weight

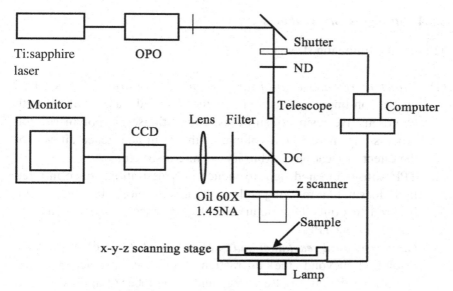

Figure 3.27: Experimental setup for 3D microfabrication via TPP.

(monomers) into macromolecules consisting of repeating units, by using light as the reaction trigger. In a TPP process, electronic transition occurs by simultaneous absorption of two photons, so that radical formation and subsequent polymerisation occurs only in the vicinity of the focused light source (laser beam). This results in a small solidified volume (voxel) around the focal spot. By either scanning the laser beam through the resin volume or by moving the sample in three perpendicular directions (X, Y and Z), arbitrary 3D microstructures can be created.

In addition, by changing the monomer or the mixture of monomers used in the resin, the characteristics of the final polymer can be changed. This makes it possible to tailor some properties of the produced microstructure. For example, polymer hardness and the amount of shrinkage the material undergoes due to polymerisation can be controlled. Furthermore, by doping the base resin with different compounds, it is possible to change physical, chemical, biological and optical properties of the final microstructure.

3.8.4 *Strengths and weakness*

TPP has the following strengths:

(1) *Direct fabrication of arbitrary 3D microstructures.* TPP is based on a non-linear interaction of the focused laser pulses with photosensitive resin, it can be used to initialise the polymerisation process anywhere in the volume of the resin and hence allows for the direct fabrication of arbitrary 3D micro-structures.

(2) TPP utilises infrared light to induce polymerisation. Since infrared light does not harm living cells or organisms, biophotopolymers can be used to print 3D structures in the presence of living cells and organisms via TPP.

(3) *Good structural resolution and quality.* For TPP microfabrication, a resolution beyond the diffraction limit can be achieved by controlling the laser pulse energy and the number of applied pulses. Thus, the technique provides much better structural resolution and quality than the stereolithography method.

On the other hand, TPP also has the following weaknesses:

(1) *Lack of scalability.* The technique lacks scalability, making it premature for the fabrication of large-scale architected structural materials.

(2) *Negative effects accompanied by high resolution.* To achieve maximum resolution, the energy input needs to be small. However, when the input energy is close to the minimum energy threshold, the correlations between voxel size, laser intensity and writing speed get highly nonlinear and are subjected to fluctuations easily.

3.8.5 *Potential applications*

(1) *Applications in photonics.* The unique optical properties of the polymers used are effective for the fabrication of micro-optical components and devices such as microprism arrays diffractive optical elements.

(2) *Biological applications*. TPP shows promise for biological applications, such as tissue engineering, drug delivery, medical implants and medical sensors. For tissue engineering, TPP can be used to produce arbitrary 3D scaffold structures as it allows precise control over the 3D geometry of the scaffold, modelling and reproduction of cellular microenvironment. In addition, the high resolution of the technique enables control over the cell organisation inside the scaffold and cell interactions consequently.

(3) *Microneedles*. TPP can be utilised for the fabrication of drug delivery devices, such as microneedle arrays. The flexibility of TPP enables arbitrary changes to the needle design and the improved geometry leads to better mechanical and puncturing properties.

3.9 3DCeram's Ceramic Parts

3.9.1 *Company*

Renamed 3DCeram in 2010, the company was created in 2001 by the Centre for Technology Tranfers in Ceramics (CTTC) to implement the solutions resulting from research in producing ceramic components on demand. The company was bought by Richard Gaignon and Christophe Chaput in late 2009, with the support of the investment company set up by Jean Mérienne and late Michel Rasse, founders and former presidents of the Ceric group. Both of them were sensitive to the industrial project of a French high-tech company. The company is headquartered at 27, rue du Petit Theil F-87280 Limoges, France.

3.9.2 *Products*

3DCeram offers machines, materials and part-building services for the creation of ceramic parts. *Ceramaker* (see Fig. 3.28), a 3DCeram trademark, is able to manufacture ceramic parts with high precision, extraordinary level of detail, perfect finish and complex geometries. Specifications of *Ceramaker* are summarised in Table 3.8.

Figure 3.28: The *Ceramaker* printer.
Courtesy of 3DCeram

Table 3.8: Summary specifications of *Ceramaker*.

Dimensions	1000 × 2300 × 2300 mm (W×D×H))
Tank capacity	10 litres (300×300×110)
Weight	1450 kg
Electrical requirements	220–240 VAC/50 Hz
Power	2 Kw
Light source	Laser
Pixel size	~30 µm
Wavelength UV	355 nm
Layers thickness	0.010–0.125 mm
Operating temperature environment	20–25°C
Relative humidity	50%

Courtesy of 3DCeram

3.9.3 *Process*

The ceramics additive manufacturing (CAM) is based on the upstream studies on laser stereolithography applied to ceramics conducted by

Thierry Chartier in 1998 at Science of Ceramic Processes and Surface Treatment (SPCTS) in Limoges. Based on that, 3DCeram further developed both the process and the ceramic paste to industrialise this technology and uses it in various applications. The process begins when the CAD file of the part to be produced is resized according to the shrinkage rate during sintering based on technical information provided by 3DCeram. The CAD file of the part is then reworked automatically by software to cut the piece into slices, each slice corresponding to a layer that will be printed and cured. The printing struts are created automatically to optimise the speed and quality of printing. The ceramic paste is then spread over a work plan in layers (25–100 microns thickness). Each layer is cured by a UV laser that also sticks it to the previous one. The process repeats until the final part is obtained. The parts are then heat treated (debinding then sintering) in order to eliminate the resin and densify the ceramic to 100%.

3.9.4 *Principle*

Ceramaker uses photosensitive resin mixed with ceramic particles — alumina, zirconia and hydroxyapatite. The mixture in the form of paste is polymerised in the laser printing process. The materials are called 3DMIX, which are generally formatted at ambient temperature and their specific properties are determined by the sintering process in which the materials consolidate. Alumina, zirconia and hydroxyapatite are mostly used for industry, luxury and biomedical applications respectively.

The laser stereolithography of ceramics is based fundamentally on the following principles:

(1) Parts are built from a photosensitive liquid resin that cures when exposed to a UV laser beam (similar to the photopolymerisation process in Section 3.1.4), which scans through the surface of the resin.
(2) The building is done layer by layer until the parts are finished.

3.9.5 *Strengths and weaknesses*

The main strengths of the laser stereolithography process are:

(1) *High resolution and precision.* Parts produced are in proper material and directly usable for the final applications.
(2) *No minimum quantity thresholds.* This leads to more varied applications.
(3) *Flexibility and high speed.* The process is not affected by complex shapes.
(4) *Same properties as parts made in a traditional way.* The parts produced have the same properties as parts made by traditional ways such as pressing and injection.
(5) *High efficiency.* It enables the manufacture within a short lead time of good material prototypes, unique parts and series-produced components.

On the other hand, the main weaknesses of the laser stereolithography process are:

(1) *Long post-processing time.* The processes of cleaning, thermal debinding and sintering take about 4 days before the final part is obtained.
(2) *Large shrinkage.* There is about 10%–25% shrinkage during sintering and this has to be accounted for when designing the CAD file.
(3) *No multicolour or multi-material capabilities.* This limits the usefulness of the ceramic parts produced.

3.9.6 *Applications*

The applications of 3DCeram's laser stereolithography of ceramics can be divided into different areas:

(1) *Biomedical applications.* With fast ceramic production (FCP), 3DCeram's system is able to manufacture made-to-measure or small series of bone substitutes and cranial or jawbone implants.

(2) *Luxury application.* With Cerampilot, a service that combines the advantages of ceramics and rapid prototyping, the system can be used to produce any shape of watches, jewellery and other luxury items.

(3) *Industry applications.* The exceptional mechanical, magnetic, thermal, chemical and electrical properties of ceramics make it a good material for prototypes for design, analysis, verification and functional testing in industries.

3.9.7 *Research and development*

As one of the leading companies in AM of ceramics, 3DCeram has made great improvements.

During Euromold 2015, 3DCeram announced that new innovations were introduced to *Ceramaker* to simplify the loading of the ceramic paste and paste feeding of the manufacturing area. The optimisation significantly reduces ceramic paste consumption and improves production quality.

3DCeram has added ceramic slurries to its range of 3DMix pastes. These slurries are compatible with a more accessible range of desktop 3D printers. This solution, particularly suited to the development of new products, allows users to test ceramics 3D printing independently.

3.10 Other Notable Liquid-Based AM Systems

There are several other commercial and non-commercial liquid-based AM methods that are similar in terms of technologies, principles and applications to those presented in the previous sections, but with some interesting variations. These notable liquid-based AM systems shall be discussed as follow:

Computer Modelling and Engineering Technology (CMET) Inc. was established in November 1990. Based on laser lithography technology,

CMET has four models of machines: *EQ-1, Rapid Meister ATOMm-4000, Rapid Meister 6000 II* and *Rapid Meister 3000*. Each of them serves individual designers and manufacturers with different parameters. *EQ-1* is the successor model to the *Rapid Meister 6000 II* and it claims to be the world's fastest scanner developed in-house. The company's address is CMET Inc., Sumitomo Fudosan Shin Yokohama Building, 2-5-5 Shin-yokohama Kohoku-ku Yokohama, Kanagawa 222-0033, Japan.

EnvisionTEC GmbH is the same company that developed the *Prefactory*® System discussed in Section 3.4. Details of the company can be found in Section 3.4.1. Similar to RegenHU's 3D bioprinters, EnvisionTEC's Bioplotter utilises computer-aided tissue engineering. The Developer and the Manufacturer Series are designed for basic research and production, respectively.

Lithoz's lithography-based ceramic manufacturing process uses an LED light source to cure a photopolymer filled with ceramic to produce ceramic parts. *CeraFab 7500* and the new *CeraFab 8500* can produce high quality and fully functional high-perfomance parts. The secondary post-process produces 100% dense ceramic parts. The company is located at Mollardgasse 85a/2/64-69 1060, Wien, Austria.

References

1. 3D Systems Product brochure. (2014). SLA Production series.

2. 3D Systems Product brochure. (2014). ProJet® 6000, 7000, 3500, 3510, 5000 and 5500X.

3. Jacobs, PF (1992). *Rapid Prototyping & Manufacturing, Fundamentals of Stereolithography*, 11–18. Society of Manufacturing Engineers.

4. Wilson, JE (1974). *Radiation Chemistry of Monomers, Polymers, and Plastics*. New York: Marcel Dekker.

5. Lawson, K (1994). UV/EB Curing in North America, *Proceedings of the International UV/EB Processing Conference*, Florida, USA, May 1-5, 1.

6. Reiser, A (1989). *Photosensitive Polymers*. New York: John Wiley.

7. Jacobs, PF (1992). *Rapid Prototyping & Manufacturing, Fundamentals of Stereolithography*, 25–32. Society of Manufacturing Engineers.

8. Jacobs, PF (1996). *Stereolithography and other RP&M Technologies*, 29–35. Society of Manufacturing Engineers.

9. Jacobs, PF (1992). *Rapid Prototyping & Manufacturing, Fundamentals of Stereolithography*, 53–56. Society of Manufacturing Engineers.

10. Jacobs, PF (1992). *Rapid Prototyping & Manufacturing, Fundamentals of Stereolithography*, 60–78. Society of Manufacturing Engineers.

11. *3D Systems Corporation Joins White House's Manufacturing Initiative*. (Mar 2014). http://www.thestreet.com/story/12529118/1/3d-systems-corporation-joins-white-houses-manufacturing-initiative.html

12. Molitch-Hou, M (Dec 2013). *3D Systems Acquires Portion of Xerox R&D Division*. http://3dprintingindustry.com/2013/12/19/3d-systems-acquires-portion-xerox-rd-division/.

13. Durham, M (2003). Rapid prototyping — Stereolithography, selective laser sintering, and polyjet, *Advance Materials & Processes*, 161, 40–42.

14. Cheng, YL, and SJ Chen (2006). Manufacturing of cardiac models through rapid prototyping technology for surgery planning, *Materials Science Forum*, 505–507, 1063–1068.

15. Brajlih, T, I Drstvensek, M Kovacic and J Balic (2006). Optimising scale factors of the PolyJet™ rapid prototyping procedure by generic programming, *Journal of Achievement in Materials and Manufacturing Engineering*, 16, 101–106.

16. Hendrik, J. *Perfactory® — A Rapid Prototyping System on the Way to the "Personal Factory" for the End User*. Envision Technologies GmbH.

17. Stampfl, J, R Cano Vives, S Seidler, R Liska, F Schwager, H Gruber, A Wöβ and P Fratzl (2003). *Proceedings of the 1st International Conference on Advanced Research in Virtual and Rapid Prototyping*, 1–4, 659–666. Leiria, Portugal..

18. Grunewald, SJ (Mar 2014). *EnvisionTEC Launches its Latest Micro 3D Printer*. http://3dprintingindustry.com/2014/03/20/3d-printer-micro-envisiontec/

19. Zhang, W, MC Leu, Z Ji, and Y Yan (1999). Rapid freezing prototyping with water. *Materials and Design*, 20 (June), 139–145.

20. Leu, MC, W Zhang and G Sui (2000). An experimental and analytical study of ice part fabrication with rapid freeze prototyping. *Annals of the CIRP*, 49, 147–150.

21. Sui, G and MC Leu (2003). Thermal analysis of ice wall built by rapid freeze prototyping. *ASME Journal of Manufacturing Science and Engineering*, 125 (November), 824–834.

22. Bryant, FD and MC Leu (2004). Study on incorporating support material in rapid freeze prototyping. *Proceedings of Solid Freeform Fabrication Symposium* (August), 2–4.

23. Leu, MC, Q Liu and FD Bryant (2003). Study of part geometric features and support materials in rapid freeze prototyping. *Annals of the CIRP*, 52, 185–188.

24. Sui, G and MC Leu (2003). Investigation of layer thickness and surface roughness in rapid freeze prototyping, *ASME Journal of Manufacturing Science and Engineering*, 125 (August), 556–563.

25. Liu Q and MC Leu (2006). Investigation of interface agent for investment casting with ice patterns. *ASME Journal of Manufacturing Science and Engineering*, 128 (May), 554–562.

26. Wu, S, J Serbin and M Gu (2006). Two-photon polymerisation for three-dimensional micro-fabrication. *Journal of Photochemistry and Photobiology A: Chemistry*, 181(1), 1–11.

Problems

1. Describe the process flow of the 3D Systems SLA.

2. Describe the process flow of the Stratasys' Polyjet systems.

3. Compare and contrast the laser-based stereolithography systems and the Stratasys' Polyjet systems. What are the strengths for each of the systems?

4. Describe the main differences between Perfactory and other commercial AM systems which use UV curing process.

5. Which liquid-based machine has the largest work volume? Which has the smallest?

6. What is DMD, as found in Perfactory?

7. Which are the key materials for RegenHU's bioprinting?

8. As opposed to many of the liquid-based AM systems which use a photosensitive polymer, water is used in RFP. What are the pros and cons of using water?

9. How is the support of the ice part removed from the actual part?

10. Discuss the principle behind the two-laser-beams method. What are the major problems in this method?

11. What are the major materials used in ceramic printing? And what are the limitations of this process?

12. Describe the system and process for Optomec's Aerosol Jet system. List a few possible applications with the availability of Aerosol Jet technology.

CHAPTER 4

SOLID-BASED ADDITIVE MANUFACTURING SYSTEMS

Solid-based additive manufacturing (AM) systems utilise solids as the primary medium to create the part or prototype. As such, they are very different from the liquid-based photo-curing systems described in Chap. 3. They are also different from one another in that the primary form of solid materials in some systems may come as filaments or wires, some as sheets or rolls while others may be as pellets. A special group of solid-based AM systems that uses powder as the medium will be covered separately in Chap. 5.

4.1 Stratasys' Fused Deposition Modelling

4.1.1 *Company*

Stratasys Ltd. was founded in 1989 in Delaware and developed the company's AM systems based on Fused Deposition Modelling (FDM) technology. The technology was first developed by Scott Crump in 1988 and the patent was awarded in the USA in 1992. FDM uses an extrusion process to build three-dimensional (3D) models. Stratasys introduced its first AM machine, the 3D Modeler®, in early 1992. Stratasys' headquarters are located at 7665 Commerce Way, Eden Prairie, MN 55344-2080, USA.

4.1.2 *Products*

Stratasys manufactures 3D printing equipment and materials that create physical objects directly from digital data. Its systems range from affordable desktop 3D printers to large and advanced 3D production systems, making 3D printing more accessible than ever.

4.1.2.1 *Idea series*

Stratasys' Idea Series 3D printers are compact, light and affordable desktop 3D printers that enhance users' design capability at the push of a button. With FDM technology, they liberate customers' creativity and accelerate the design process. Stratasys' Mojo is shown in Fig. 4.1 and the uPrint SE Plus is shown in Fig. 4.2. These 3D printers bring professional 3D printing to consumers' desktops or small team workspaces. Details of the Idea Series 3D Printers are summarised in Table 4.1.

Figure 4.1: Mojo.   Figure 4.2: uPrint SE Plus.
Courtesy of Stratasys Ltd.   Courtesy of Stratasys Ltd.

4.1.2.2 *Design series — Performance 3D printers*

Stratasys' Design Series 3D printers are built as affordable and office-friendly AM systems. They are meant primarily for concept modelling, creating product replicas and some functional testing capabilities. Under the Design Series, there are two categories of 3D printers — Precision 3D printers and Performance 3D printers. Precision 3D printers are based on PolyJet 3D printing technology, which is discussed in Section 3.2. On the other hand, the three machines in the Performance 3D printers series — Dimension 1200es, Dimension Elite and Fortus 250mc (Fig. 4.3) — are powered by FDM technology.

Table 4.1: Specifications of the Idea Series 3D printers.

Models	uPrint SE	uPrint SE Plus	Mojo
Model materials	ABSplus in ivory	ABSplus in ivory, white, blue, fluorescent yellow, black, red, nectarine, olive green or grey	
Support materials	SR-30 soluble		
Maximum part size, mm (in)	203 × 152 × 152 (8 × 6 × 6)		127 × 127 × 127 (5 × 5 × 5)
Layer thickness, mm (in)	0.254 (0.010)	0.254 (0.010) or 0.330 (0.013)	0.17 (0.007)
Size, mm (in)	With one material bay 635 × 660 × 787 (25 × 26 × 31) With two material bays 635 × 660 × 940 (25 × 26 × 37)		630 × 450 × 530 (25 × 18 × 21)
Weight, kg (lbs)	With one material bay 76 (168) With two material bays 94 (206)		27 (60)
Software	CatalystEX software		Print Wizard software

Figure 4.3: Stratasys' performance 3D printers, Fortus 250mc (left), dimension 1200es (centre), dimension elite (right).
Courtesy of Stratasys Ltd.

Performance 3D printers use ABSplus thermoplastic to produce the 3D models. The parts are durable and dimensionally stable, making them perfect for tough testing. The raw materials are affordable, allowing frequent iterative 3D modelling. Details of the Design Series Performance 3D printers are summarised in Table 4.2.

Table 4.2: Specifications of the design series — Performance 3D printers.

Models	Fortus 250mc	Dimension 1200es	Dimension Elite
Build size, mm (in)	254 × 254 × 305 (10 × 10 × 12)		203 × 203 × 305 (8 × 8 × 12)
Materials	ABSplus-P430	ABS or ABSplus plastic in standard natural, blue, fluorescent yellow, black, red, olive green, nectarine and grey colours Custom colours are available (Dimension Elite uses only ABSplus)	
Support structures	Soluble support	Breakaway support or soluble supports	Soluble support
Material cartridges	One autoload cartridge with 922 cm^3 (56.3 in^3) build material One autoload cartridge with 922 cm^3 (56.3 in^3) support material		
Layer thickness, mm (in)	0.330 or 0.254 or 0.178 (0.013 or 0.010 or 0.007)	0.254 or 0.33 (0.010 or 0.013)	0.178 or 0.254 (0.007 or 0.010)
Size, mm (in)	838 × 737 × 1143 (39 × 29 × 45)		686 × 914 × 1041 (27 × 36 × 41)
Weight, kg (lbs)	148 (326)		136 (300)
Software	Insight software	CatalystEX software	

4.1.2.3   *Production series*

Stratasys' Production Series 3D printers can create large products on the factory floor, having a build size larger than the office-friendly Performance Series or the desktop Idea Series. They can create low-volume assembly fixtures and jigs directly from computer-aided design (CAD) data. Thus they can be used for prototyping, tooling and digital manufacturing for designers and engineers. There are three FDM machines in the Production Series, Fortus 380mc, Fortus 450mc and Fortus 900mc (Fig. 4.4). The Fortus 900mc 3D Production System offers the use of the widest range of FDM materials. They include high-performance thermoplastics for durable and accurate parts as large as

Figure 4.4: Stratasys' Production Series 3D printers — Fortus 450mc (left), Fortus 900mc (right).
Courtesy of Stratasys Ltd.

Table 4.3: Specifications of the Production Series 3D printers.

Models	Fortus 380mc/450mc	Fortus 900mc
Model materials	ABS-M30, ABS-M30i, ABS-EDS7, ASA, PC-ISO in white and translucent, PC, FDM Nylon 12, (Fortus 450mc only) ULTEM 9085 resin, ULTEM 1010 resin	ABS-ESD7, ABSi, ABS-M30, ABS-M30i, ASA, FDM Nylon 12, PC, PC-ABS, PC-ISO, PPSF, ULTEM 9085 resin, ULTEM 1010 resin
Build envelope, mm (in)	Fortus 380mc: 355 × 305 × 305 (14 × 12 × 12) Fortus 450mc: 406 × 355 × 406 (16 × 14 × 16)	914 × 610 × 914 (36 × 24 × 36)
Layer thicknesses, mm (in)	0.127, 0.178, 0.254, 0.330 (0.005, 0.007, 0.010, 0.013)	0.178, 0.254, 0.330 (0.007, 0.010, 0.013)
Support structure	Soluble for most materials; break-away for PC-ISO and ULTEM; soluble or break-away for PC	Soluble for most materials; break-away for PC-ISO, ULTEM and PPSF; soluble or break-away for PC
Size, mm (in)	1295 × 902 × 1984 (51 × 35.5 × 78.1)	2772 × 1683 × 2027 (109.1 × 66.3 × 79.8)
Weight, kg (lbs)	601 (1325)	3287 (7247)
Software	Insight software	

914 × 610 × 914 mm (36 × 24 × 36 in) with strong mechanical, chemical and thermal properties. Details of the Production Series 3D printers are summarised in Table 4.3.

4.1.2.4 *Materials*

Stratasys Ltd. offers a wide range of AM materials, including clear, rubberlike and biocompatible photopolymers for the PolyJet machines and tough high-performance thermoplastics for the FDM machines. The wide range of materials enables designers and engineers to build practically anything at any stage of the product development cycle, from fast and affordable concept modelling, to detailed, realistic and functional prototyping, to certification testing, and to agile, low-risk production. Furthermore, FDM thermoplastics are used to build tough, durable parts that are accurate, repeatable and stable over time. Materials used include ABS, polycarbonate, FDM Nylon-12 and ULTEM™, a thermoplastic polyetherimide (PTI) that has good thermal resistance, high strength and stiffness, and broad chemical resistance.

4.1.3 *Process*

In Stratasys' patented process,[1] a geometric model of a conceptual design is created on CAD software which uses .STL or Initial Graphics Exchange Specification (IGES) files. It can then be exported from the CAD software into the AM systems where it is processed using Insight software or CatalystEX software, which automatically generates supports. Within this software, the CAD file is sliced into horizontal layers after the part is oriented for the optimum build position, and any necessary support structures are automatically detected and generated. The slice thickness can be set manually anywhere between 0.127 and 0.330 mm (0.005–0.013 in.) depending on the needs of the models and the machine. Tool paths of the build process are then generated and downloaded to the FDM machine.

Modelling material is in the form of a filament, very much like a stiff fishing line, and is stored in a cartridge or spool. The filament is fed into an extrusion head and heated to a semi-liquid state. The material is then extruded through the head and then deposited in ultra-thin layers from the FDM head, one layer at a time. Since the air surrounding the head is maintained at a temperature below the material's melting point, the exiting material quickly solidifies. Moving on the x-y plane, the head follows the tool path generated by the software, fabricating the desired layer. When the layer is completed, the base plate moves down one layer and the head starts creating the new layer. Two filament materials are dispensed through a dual tip mechanism in the FDM machine: a primary modeller material is used to produce the model geometry and a secondary material, or release material, is used to produce the support structures. The release material forms a bond with the primary modeller material and can be broken away or washed away with detergent and water upon completion of the 3D models. A schematic diagram of the FDM process is shown in Fig. 4.5.

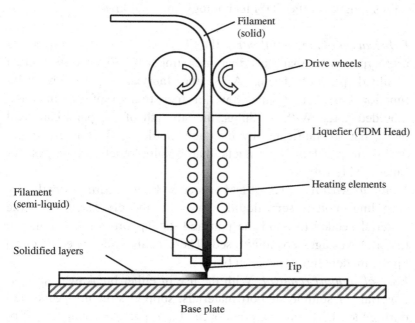

Figure 4.5: The basic FDM process.

4.1.4 *Principle*

The principle of FDM technology is based on surface chemistry, thermal energy and layer manufacturing technology. It is known for its reliability and durable parts, and extrudes fine lines of molten thermoplastic that solidify as they are deposited. 3D printers that run on FDM technology build parts layer-by-layer by heating thermoplastic material to a semi-liquid state and extruding it according to its computer-controlled paths. Parameters that can be changed include material properties (such as the material column strength, material flexural modulus and material viscosity), positioning accuracy, road widths, deposition speed, volumetric flow rate, tip diameter, envelope temperature, part geometry and orientation.

4.1.5 *Strengths and weaknesses*

The main strengths of the FDM technology are as follows:

(1) *Fabrication of functional parts*: the FDM process is able to fabricate prototypes with materials that are similar to that of the actual moulded product. Using ABS, the fabricated parts are fully functional parts and have 85% of the strength of the injection moulded parts. With ABSplus, the strength of the parts can rival those that are fabricated by injection moulding. This is especially useful in developing products that require quick prototypes for functional testing.

(2) *Minimal wastage*: The FDM process builds parts directly by extruding molten semi-liquids onto the model. Thus, only the material needed to build the part and its support is needed and so material wastages are minimised. There is also little need to clean up the model after it has been built.

(3) *Ease of support removal*: With the use of either breakaway support materials or soluble support materials, support structures generated during the FDM building process can be respectively snapped off or simply washed away in a water-based solution. This makes it very

convenient for users to get to their finished parts very quickly and there is very little or no post-processing necessary.

(4) *Ease of material change*: Build materials, supplied in spool or cartridge form, are easy to handle and can be changed readily when the materials in the system are running low. This keeps the operation of the machine simple.

(5) *Large build volume*: FDM machines, especially the Fortus 900mc, offer a larger build volume than most of the other AM systems available.

The weaknesses of the FDM technology are as follows:

(1) *Restricted accuracy*: Parts built with the FDM process usually have restricted accuracy due to the shape of the material used, that is, the filament form. Typically, the filament used has a diameter of 1.27 mm and this tends to set a limit on how accurate the part can be achieved. However the newer FDM machines have made significant improvements in mitigating this issue by using better machine control.

(2) *Slow process*: The building process is slow, as the whole cross-sectional area needs to be filled with building materials. Building speed is restricted by the extrusion rate or the flow rate of the build material from the extrusion head. As the build materials used are plastics and their viscosities are relatively high, the build process cannot be easily sped up.

(3) *Unpredictable shrinkage*: As the build material cools rapidly upon deposition from the extrusion head, there will be thermal contraction and stresses induced. As such, shrinkages and distortions in the model are common and are usually difficult to predict. With experience, users may be able to compensate for these issues by adjusting the process parameters of the machine.

4.1.6 *Applications*

FDM models can be used in the following general application areas:

(1) *Models and prototypes for conceptualisation and presentation*: FDM 3D printers can create models and prototypes for new product design and testing and build finished goods in low volumes.

(2) *Educational use*: Educators can use FDM technology to elevate research and learning in science, engineering, design and art.

(3) *Customisation of 3D models*: Hobbyists and entrepreneurs can use FDM to manufacture products in their homes — creating gifts, novelties, customised devices and inventions.

4.1.7 *Research and development*

Since its founding in 1989, Stratasys Ltd. has grown significantly over the last 25 years, acquiring Solidscape Inc. in May 2011. Stratasys also completed its merger with Objet Ltd., retaining the name Stratasys Ltd. and forming a company with capitalisation of approximately $3 billion in December 2012. MakerBot, a subsidiary of Stratasys since being acquired in June 2013, manufactures the company's prosumer desktop 3D printers in Brooklyn, New York. MakerBot maintains the Thingiverse design-sharing community and facilitates a wide network of user groups. Stratasys also operates Stratasys Direct Manufacturing, formerly known as RedEye On Demand, a digital manufacturing service. Through its network of certified resellers, Stratasys delivers responsive, regional support around the globe. The company maintains dual headquarters in Eden Prairie, Minnesota and Rehovot, Israel, and has opened up a new headquarters in Rheimmünster, Germany. Stratasys holds 800 granted or pending AM patents worldwide.[2] It is a public company that trades on NASDAQ (SSYS).

4.2 Mcor Technologies' Selective Deposition Lamination

4.2.1 *Company*

Mcor Technologies Ltd was founded in 2005 and the company's AM systems are developed based on Selective Deposition Lamination (SDL) technology, which is also known as paper 3D printing. This technology was first invented by Dr. Conor MacCormack and Fintan MacCormack in 2003. Mcor Technologies Ltd's printers are the only 3D printers to use ordinary and affordable office paper as the build material. Mcor Technologies Ltd is located at Unit 1, IDA Business Park, Ardee Road, Dunleer, Co Louth, Ireland.

4.2.2 *Products*

Mcor Technologies Ltd aims to produce eco-friendly and user-friendly paper-based 3D printers with lower cost. There are three 3D printers: Mcor Matrix 300+, Mcor IRIS HD and Mcor ARKe.

Mcor Matrix 300+ uses A4-sized or letter-sized paper and a water-based adhesive to produce models at a low cost. Its printing speed has been improved up to 3 times.

Mcor IRIS HD is a colour 3D printer that can produce high quality (5760 × 1440 × 508 dpi) coloured products with more than 1 million colours. The full colour printing is ensured by Mcor 3D Ink. Mcor IRIS HD is an affordable 3D printer that is suitable for use in offices due to its small size and low noise level.

MCor ARKe was launched in January 2016. It bundles in Mcor Mobile App and Mcor Orange software for Mac and Windows PCs for easy sharing and monitoring of printing projects. It operates four times faster than the IRIS HD. The ARKe utilises a roll of paper rather than sheets of paper, as with the previous two models.

Details of all three printers are summarised in Table 4.4. Illustrations of the printers are shown in Fig. 4.6.

ColourIT is a software developed by Mcor Technologies Ltd to enable users to add colours directly to the surfaces of their 3D CAD model. It is compatible with most file formats, including .STL, .WRL, .OBJ, .3DS, .FBX, .DAE and .PLY. After adding the colours, the CAD model is

Figure 4.6: The Mcor Matrix 300+ (left), IRIS HD (centre) and ARKe (right) 3D printers. Courtesy of Mcor Technologies Ltd.

Table 4.4: Specifications of Mcor Matrix 300+, IRIS and ARKe 3D printers.

Models	Mcor Matrix 300+	Mcor IRIS HD	Mcor ARKe
Resolution	0.1 mm (0.004 in)	Colour resolution x × y × z: 5760 × 1440 × 508 dpi Axis resolution x × y × z: 12 × 12 × 100 μm (0.0004 × 0.0004 × 0.004 in)	Colour resolution x × y × z: 4800 × 2400 × 254 (or up to 508 with 50 gsm) dpi Axis resolution: 0.1 mm (0.004 in)
Build size	A4: 256 × 169 × 150 mm Letter: 9.39 × 6.89 × 5.9 in		240 × 205 × 125 mm (9.5 × 8 × 4.9 in)
Build material	A4 (80 gsm, 160 gsm) or Letter (20 lbs, 43 lbs)	A4 (80 gsm, 160 gsm ply colour only) or Letter (20 lbs, 43 lbs ply colour only)	Rolls of paper
Layer thickness, mm (in)	0.1 (0.004), 0.19 (0.007) in ply colour only		nil
File formats	.STL, .OBJ, VRML, COLLADA		.STL, .OBJ, VRML, .DAE, .3MF
Equipment dimensions, mm (in)	950 × 700 × 800 (37.4 × 27.55 × 31.5)		880 × 593 × 633 (34.6 × 23.3 × 24.9)

exported by ColourIT as a .WRL file, which can be imported to the SliceIT software on Mcor IRIS HD printers.

Mcor Orange software and Mcor Mobile app are available for use in the Mcor ARKe for easy sharing of designs and monitoring and managing of printing projects.

4.2.3 *Process*

SDL is primarily a "lamination and cut" process which can be described as follows:

(1) A .STL or .OBJ or VRML (for colour 3D printing) file is imported into Mcor Technologies Ltd's 3D printers.

(2) With Mcor Technologies Ltd's control software, SliceIT, the CAD model is sliced into printable layers equivalent in thickness to the standard papers. The additional software, ColourIT, is used to add colours onto the CAD model.

(3) After the slicing file is ready, the first sheet of paper is attached to the build plate.

(4) The 3D printer deposits drops of adhesive onto the first sheet of paper according to the sliced cross-section.

(5) A new sheet of paper slides in and pressure is exerted by the printer to bond these two sheets together.

(6) An adjustable tungsten carbide blade cuts one sheet of paper at a time, tracing the object outline and necessary cross-hatches to create the edges of the part.

(7) Steps 3–6 are repeated as the build platform lowers, until the last sheet of paper is added.

(8) The solid object is then removed manually from the surrounding paper

To print in colour, a standard Epson 2D colour inkjet pre-prints duplex colour onto each sheet, using a special ink designed to soak into the page. Only the edges of the model are printed with colour. Each sheet is also printed with a barcode to verify the correct sequence of sheets.

4.2.4 *Principle*

The principle of SDL is based on the simple addition of sheets of paper. This process is very similar to one which Cubic Technologies used to market as Laminated Object Manufacturing (LOM) except that rather than using a roll of adhesive-coated paper and a laser cutter as in the LOM, standard sheets of paper (A4 sized or letter sized) together with an adhesive dispensing system and tungsten-tip knife are used. Various sheets of papers are accumulated with the adhesive to form the final product and an adjustable tungsten carbide blade will cut the sheets one at a time, tracing the object outline to create the edges of the part. Outside the edges, cross-hatches are created to facilitate removal at the end. Hence, the final product is produced when all the waste is removed. Schematics of the SDL and LOM (which is no longer sold) processes are shown in Figs. 4.7 and 4.8, respectively.

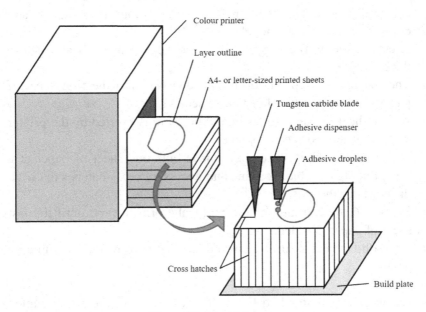

Figure 4.7: The SDL process.

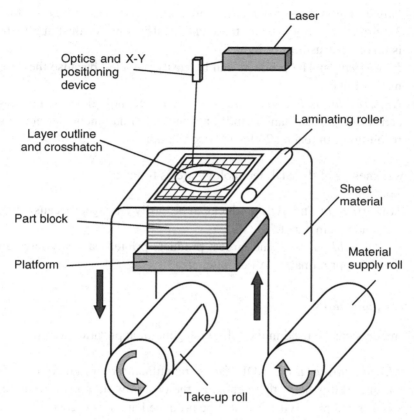

Figure 4.8: The LOM process.

4.2.5 *Strengths and weaknesses*

The main strengths of the SDL technology are as follows:

(1) *Low cost*: Mcor Technologies' 3D printers produce parts at possibly the lowest price in the industry — approximately 5% of other similar technologies' costs, because the technology mainly uses stacks of papers as its building material.
(2) *High precision*: Mcor Technologies' 3D printers can build models to a precision of 0.00047 in (0.012 mm) and a dimensional accuracy of 0.004 in (0.1 mm).

(3) *Safety*: Throughout the making process, Mcor Technologies Ltd's 3D printers do not use any toxic chemicals, fumes or dust and there is no dangerous heat or light.
(4) *Eco-Friendly*: The products can be easily recycled because they are made of paper.
(5) *High resolution in colour printing*: Mcor Technologies' 3D printers can print more than 1 million colours simultaneously and its resolution can reach 5760 × 1440 × 508 dpi.

The weaknesses of the SDL technology are as follows:

(1) *Low strength*: The SDL product can be easily damaged because they are made from paper.
(2) *Small build volume*: The largest products printed can only have an area of approximately A4 or letter size.

4.2.6 *Applications*

SDL models can be used in the following general application areas:

(1) *Manufacturing uses*: SDL helps manufacturers produce models faster and this helps them enhance the design within a shorter time. Conceptual prototypes can also be produced at a lower cost.
(2) *Architectural uses*: SDL produces more accurate and precise models or prototypes for architects. This helps to save time and cost for them and it also reduces architects' working period.
(3) *Marketing*: Models made by SDL are more attractive and powerful in showcasing prospects to customers before the actual products are launched in the market.

4.3 Sciaky's Electron Beam Additive Manufacturing

4.3.1 *Company*

Sciaky, Inc. was established in 1939 as a welding services workshop and provider of advanced welding systems in the fabrication market. In 1994, it was acquired by Phillips Service Industries, Inc. (PSI). A year later Sciaky began research on a new AM process using metal wire feedstock and electron beams. In 2009 this became known as Electron Beam Additive Manufacturing (EBAM) and Sciaky started to offer EBAM as a service option. In 2014 Sciaky started to sell EBAM turnkey systems in the commercial market. Sciaky, Inc. is headquartered at 4915 W 67th St., Chicago, IL 60638, USA.

On top of its EBAM systems, Sciaky still offers contract AM services as well as electron beam and arc welding services.

4.3.2　Products

Sciaky offers five turnkey EBAM systems, differing only in their build chamber size, work envelope and part envelope. EBAM systems will reduce material costs and material wastage, shorten machining times, and reduce the need for post-processing machining. They are also faster than other metal AM systems, which mostly use powder as their starting material. Specifications of the EBAM series and an illustration of one of the systems are given in Table 4.5 and Fig. 4.9, respectively.

Figure 4.9: Sciaky, Inc.'s EBAM 300.
Courtesy of Sciaky, Inc.

Table 4.5: Specifications of the EBAM series.

Models	EBAM 300	EBAM 150	EBAM 110	EBAM 88	EBAM 68
Chamber dimensions, mm (in)	7620 × 2743 × 3353 (300 × 108 × 132)	3810 × 3810 × 3048 (150 × 150 × 120)	2794 × 2794 × 2794 (110 × 110 × 110)	2235 × 2235 × 2794 (88 × 88 × 110)	1727 × 1727 × 2794 (68 × 68 × 110)
Work envelope, mm (in)	5791 × 1219 × 1219 (228 × 48 × 48)	2794 × 1575 × 1575 (110 × 62 × 62)	1778 × 1194 × 1600 (70 × 47 × 63)	1219 × 889 × 1600 (48 × 35 × 63)	711 × 635 × 1600 (28 × 25 × 63)
Nominal part envelope, mm (in)	7112 × 1219 × 1219 (280 × 48 × 48)	3708 × 1575 × 1575 (146 × 62 × 62)	2692 × 1194 × 1600 (106 × 47 × 63)	2134 × 889 × 1600 (84 × 35 × 63)	1625 × 635 × 1600 (64 × 25 × 63)

All models in the series have a high efficiency pumping-chamber hard vacuum of 1×10^{-4} Torr. Deposition rates range from 3 to 9 kg per hour, depending on part geometry and material used. Through the use of a custom real-time camera system, the patented closed-loop control (CLC) technology automates the EBAM process and constantly monitors and adjusts parameters to maintain the appropriate molten pool size.

Sciaky recommends EBAM machines to use weldable metals available in wire feedstock form, including titanium, titanium alloys, Inconel 718, Inconel 625, tantalum, tungsten, niobium, stainless steels, aluminium, zirconium alloys and copper-nickel alloys. EBAM machines can use wire sizes of 0.9 mm, 1.1 mm, 1.6 mm, 2.4 mm, 3.2 mm and 4.0 mm.

4.3.3 *Process*

An EBAM machine contains an EBAM gun and a platform inside a vacuum chamber. The EBAM gun consists of the electron beam and the dual wire feed system which can contain wires made of two different metals. The EB gun can move in the y- and z-directions, while the platform moves in the x-direction.

To start off, the EBAM machine slices the 3D model and converts them into computer numerical control (CNC) code. The wire feeder deposits wire(s) onto the substrate. The electron beam focuses onto the newly deposited wire, melting it onto the surface as in Fig. 4.10. As the platform moves in the x-direction away from the electron beam, the metal solidifies back to solid form.

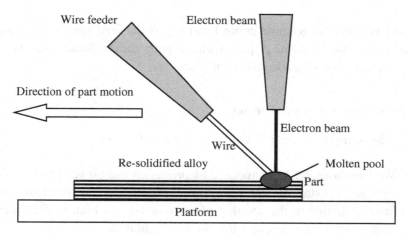

Figure 4.10: The EBAM process.

Once the part reaches near-net shape, it undergoes heat treatment and machining.

EBAM machines also have the ability to switch between thin wires and thick wires for fine and coarse deposition features respectively. Gross deposition rates vary from 3 to 9 kg per hour, depending on material and part features.

4.3.4 *Principle*

When a powerful electron beam is focused onto a piece of metal wire, the wire melts and forms a molten pool at the surface of the part. The size of this molten pool is indicative of the metal's microstructure and mechanical properties, as well as the layer geometry and metal chemistry. The molten pool is monitored by a camera in a patented CLC system called the Interlayer Real-time Imaging and Sensing System (IRISS), which automatically adjusts the processing parameters, such as electron beam power, CNC motion profiles and wire feed rate(s) to maintain molten pool size. Once the electron beam moves away, the molten pool solidifies.

An independent programme control can adjust the feed rates of the two different metals to create a part with varying compositions, that is, a "custom" alloy or a "graded material" part.

4.3.5 *Strengths and weaknesses*

The main strengths of the EBAM process are as follows:

(1) *Nonflammability*: Most metal AM processes use powder instead of wire as their starting material. Powder feedstocks can sometimes be flammable due to the aeration of the powder and high surface area contact with air, whereas wires are not flammable.
(2) *Low material wastage*: Powder feedstocks use a large vat of powder, which means that there will be material left over after the printing. Wires do not have this problem.
(3) *Low cost*: Wire feedstocks are generally less expensive than powder feedstocks.
(4) *Varying material composition*: With the proprietary dual wire feed system, EBAM allows the mixing of two different metals in varying proportions in the printing of the part.
(5) *Switching between fine and coarse features*: EBAM machines allow for the switching between thin and thick wires to either print fine features or to print parts quickly, thus improving the production rate.
(6) *High deposition rates*: Depending on material and part features, deposition rates range between 3 and 9 kg per hour, making it one of the fastest metal directed energy deposition processes.

The main weaknesses of the EBAM process are as follows:

(1) *Inability to print overhangs*: EBAM machines are not able to print any supports, which is a weakness inherent to most solid AM processes.
(2) *Poor resolution*: EBAM cannot print fine detail features limited by the wire diameter and will therefore require machining to achieve good surface finish. Due to this requirement, it is difficult to fabricate parts with complex internal geometries.
(3) *Limited materials*: EBAM machines can only use weldable metals as starting materials.

4.3.6 *Research and development*

Sciaky, Inc. conducts several military and aerospace research programmes. For instance Sciaky received federal funds to work with the Air Force Research Lab under their Small Business Innovative Research (SBIR) programme to serve the technology and cost needs of the U.S. Air Force.

Sciaky has received an SBIR award for refining and demonstrating CLC in EBAM, with the aim to optimise control during the deposition of a molten titanium pool to ensure repeatability of the process for the production of components for the aerospace industry.[3] Sciaky has also received an SBIR award for demonstrating from start to finish the processes and procedures involved in providing cheaper titanium components using EBAM for the aerospace industry that meet the quality and affordability requirements of the U.S. Air Force.[4]

Through Defense Advanced Research Projects Agency (DARPA) funding, Sciaky, Inc. has teamed up with Pennsylvania State University's Applied Research Laboratory to create a Centre for Innovative Metal Processing through Direct Digital Deposition (CIMP-3D).[5] The Centre hopes to successfully implement metal 3D printing for critical metallic components and structures and to promote the use of such technology through training, education and dissemination of information.

4.4   Fabrisonic's Ultrasonic Additive Manufacturing

Ultrasonic additive manufacturing (UAM), previously known as ultrasonic consolidation (UC), is a solid-state metal AM process in which metal foils are bonded together using ultrasonic welding to produce three dimensional objects.[6] CNC milling is sequentially used to introduce features that are designed in the part and for the final finishing of the part. Therefore, the process combines both additive and subtractive manufacturing to obtain a desired part. One of the key features that makes UAM distinct from other metal AM processes like EBAM is that it does not involve melting. The temperatures involved at the mating interfaces are less than half of the melting point of the materials.[7] This makes the process suitable for some important applications such as embedding electronics, sensors and optical fibres.

4.4.1   *Company*

Fabrisonic, LLC was formed in 2011 as a joint venture between EWI, a non-profit engineering consulting firm, and Solidica, Inc., the original developer and manufacturer of UAM systems. Solidica, Inc. sold 15 UAM systems from 2001 to 2010, and Fabrisonic has sold five UAM systems from 2011 to the end of 2015. Fabrisonic continues to offer its UAM systems although it has moved to a parts sales and parts production model. Fabrisonic and EWI are located at 1250 Arthus E. Adams Drive, Columbus, OH 43221-3585, USA, near Ohio State University.

4.4.2   *Principle*

The UAM process utilises a cylindrical sonotrode to produce high frequency (20 kHz) and low-amplitude (20–50 µm) mechanical vibrations to induce shear with the two mating surfaces of two metallic foils held together by normal forces. The high frequency vibration induces localised friction and enables the breakage of oxide films at the contact surfaces of the foils. The oxide layer debris and any other contaminants between the foils are flushed out due to vibration, creating two nascent surfaces. When the two nascent surfaces are brought into atomic level contact by the application of a normal force and by the localised heat generation due to friction, a metallurgical bond is realised. The bonding mechanism at the interface between two foils is due to the localised plastic deformation of the asperities and grain growth across the interface.[8]

To fabricate a 3D part, a thin metallic foil (100–150 μm) is held firmly on a substrate plate either by mechanical clamping or by pneumatically assisted clamping. A sonotrode is then traversed over the foil at a certain amplitude and frequency. The metal foil also vibrates in ultrasonic frequency, resulting in a proper bonding of the metal foil to the substrate. Then a second foil is placed over the already deposited foil and the process of sonotrode traverse is repeated. As the layer-by-layer addition of the foil progresses, a CNC milling head is brought in to create any features designed on the part or to finish the surface or contours. Depending on the nature of material and to reduce the effect of any residual stresses, the base plate is pre-heated to an elevated temperature. Fig. 4.11 shows the schematic of the working principles of the UAM process.

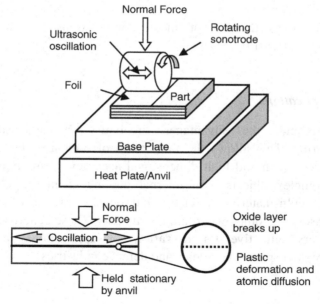

Figure 4.11: Schematic representation of the UAM process.

4.4.3 *Process parameters*

UAM process parameters have a direct consequence on the quality of the final part. The important parameters in UAM are the ultrasonic amplitude (approximately 5–50 µm), the applied normal force (500–2000 N), the travel speed of the sonotrode horn (up to 50 mm/s), the texture of the sonotrode horn (R_a between 4 and 15 µm) and the substrate preheat temperature (up to 150°C). These parameters vary for different materials and need to be optimised for every material.[9] The placement of foils and its orientation also play a major role in deciding the final properties of the part.

A commonly used parameter to represent the bond quality in UAM is linear weld density (LWD), which is given by

LWD = Bonded interface length/Total interface length

Unbonded interfaces will show up as voids within the metal, parallel to the layering.

Table 4.6 shows the list of metal combinations that have been successfully tested in UAM.

4.4.4 *Applications*

UAM combines the advantages of both AM and subtractive manufacturing. This allows the fabrication of complex 3D parts with good surface finish and high dimensional accuracy. Some typical parts include complex objects with internal cooling channels, honeycomb structures, multi-material laminates and objects integrated with fibre optics, sensors and other electronic instruments. These benefits make the process very attractive for a variety of applications in tooling, automotive, aerospace, electronics and defence industries.

Table 4.6: List of metal combinations proven to work in UAM.

	Al	Be	Cu	Ge	Au	Fe	Mg	Mo	Ni	Pd	Pt	Si	Ag	Ta	Sn	Ti	W	Zr
Al alloys	✓	✓	✓	✓	✓	✓	✓	✓	✓	✓	✓	✓	✓	✓	✓	✓	✓	✓
Be alloys		✓	✓	✓								✓						
Cu alloys			✓		✓	✓	✓	✓	✓	✓			✓	✓		✓	✓	✓
Ge					✓						✓							
Au					✓	✓	✓	✓	✓	✓				✓		✓		✓
Fe alloys						✓	✓	✓					✓	✓		✓	✓	✓
Mg alloys							✓				✓			✓				
Mo alloys								✓	✓		✓		✓			✓	✓	✓
Ni alloys									✓	✓	✓		✓			✓	✓	
Pd										✓	✓	✓						
Pt alloys											✓	✓		✓		✓	✓	
Si												✓	✓					
Ag alloys													✓	✓				✓
Ta alloys														✓		✓	✓	
Sn															✓			
Ti alloys																✓	✓	
W alloys																	✓	
Zr alloys																		✓

✓ Material pair proven to work in UAM

4.5 Other Notable Solid-Based AM Systems

There are several other commercial and research solid-based AM methods. Some of these AM systems (such as CAM-LEM Inc.'s CL-100) have not been commercialised and are still being researched, while others are no longer available in the market due to the advancement of new systems and market competition. Some of the systems that have been taken off the market include Cubic Technologies' LOM, Solidimension's Plastic Sheet Lamination (PSL) and Kira Corporation's Paper Lamination Technology (PLT).

4.5.1 *Solidscape's 3Z series*

Solidscape, Inc. was acquired in May 2011 by Stratasys and functions as a wholly owned subsidiary of Stratasys. Solidscape has sold over 4000 systems in 80 countries. Today, Solidscape sells four 3D printers that makes wax patterns ideal for lost wax investment and mould making applications — the Solidscape MAX^2, Pro, Studio and Lab. A Solidscape Pro printer is shown in Fig. 4.12. These printers are based on the FDM principle of Stratasys' machines, except that the materials are jetted via droplets, and a milling head will ensure the flatness of each layer before the deposition of the next layer. The company is headquartered at 316 Daniel Webster Highway, Merrimack, NH 03054, USA.

Figure 4.12: Solidscape Pro with touch screen.
Courtesy of Solidscape, Inc.

References

1. Crump, S (1992). The extrusion of fused deposition modeling, *Proceedings of 3rd International Conference on Rapid Prototyping*, pp. 91–100.

2. Stratasys Inc. (2016). *STRATASYS launches World's First Full-Color Multi-material, the j750 3d printer, in Singapore.* Retrieved from http://www.stratasys.com/corporate/newsroom/press-releases/asia-pacific/may-10-2016

3. Air Force SBIR/STTR. Contract FA8650-11-C-5165, Award Details. Retrieved from https://www.afsbirsttr.com/award/AwardDetails.aspx?pk=18447

4. Air Force SBIR/STTR. Contract FA8650-11-C-5302, Award Details. Retrieved from https://www.afsbirsttr.com/award/AwardDetails.aspx?pk=18812

5. Penn State Applied Research Laboratory. Materials & Manufacturing, Center for Innovative Metal Processing through Direct Digital Deposition (CIMP-3D). Retrieved from https://www.arl.psu.edu/mm_lp_cimp3d.php

6. White, DR (2003). Ultrasonic consolidation of aluminium tooling. *Advanced Materials and Process,* 161, 64–65.

7. Pal, D and B Stucker (2013). A study of sub-grain formation in Al 3003 H-18 foils undergoing ultrasonic additive manufacturing using a dislocation density based crystal plasticity finite element frame work. *Journal of Applied Physics,* 113, 203517; doi: 10.1063/1.4807831.

8. Dehoff, RR and SS Babu (2010). Characterization of interfacial microstructures in 3003 aluminum alloy blocks fabricated by ultrasonic additive manufacturing. *Acta Materialia,* 58, 4305–4315.

9. Janaki Ram, GD, Y Yang and BE Stucker (2006). Effect of process parameters on bond formation during ultrasonic consolidation of aluminum alloy 3003. *Journal of Manufacturing Systems,* 25(3), 221–238.

Problems

1. Describe the process flow of Mcor's SDL.

2. Describe the process flow of Stratasys' FDM.

3. Compare and contrast the SDL process and the FDM systems. What are the advantages and disadvantages of each system?

4. How does a metal wire melt and re-solidify in Sciaky's Electron Beam Additive Manufacturing (EBAM) process? What other processes do you think can allow this melting and re-solidifying to occur?

5. Describe the critical factors that will influence the performance and functions of:

   (i) Stratasys' FDM,

    (ii)   Mcor Technologies' SDL,

    (iii)   Sciaky's EBAM,

    (iv)   Fabrisonic's Ultrasonic Additive Manufacturing.

6.   Compare and contrast Cubic's LOM process with Mcor Technologies' SDL.

7.   What are the advantages and disadvantages of solid-based systems compared with liquid-based systems?

CHAPTER 5

POWDER-BASED ADDITIVE MANUFACTURING SYSTEMS

This chapter describes the special group of solid-based additive manufacturing (AM) systems, which primarily use powder as the basic medium for printing. Some of the systems in this group, like Selective Laser Sintering (SLS), bear similarities to the liquid-based AM systems described in Chap. 3, that is, they generally have a laser to "draw" the part layer-by-layer, but the medium used for building the model is a powder instead of photocurable resin. Others, like ColorJet Printing (CJP) has similarities with the solid-based AM systems described in Chap. 4. The common feature among these systems described in this chapter is that the material used for building the part or prototype is invariably powder based.

5.1 3D Systems' SLS

5.1.1 *Company*

3D Systems was founded by Charles W. Hull and Raymond S. Freed in 1986 commercialising the SLA systems. 3D Systems acquired DTM Corporation, the original company that first introduced the SLS® technology, in August 2001. DTM was first established in 1987. With financial support from BFGoodrich Company and based on the technology that was developed and patented at the University of Texas, Austin, USA, DTM shipped its first commercial machine in 1992. It had worldwide exclusive license to commercialise the SLS® technology till its acquisition by 3D Systems. The address of 3D Systems' head office is 333 Three D Systems Circle, Rock Hill, SC 29730, USA.

5.1.2 *Products*

3D Systems has introduced several generations of the SLS system since the acquisition of DTM. The current generation consists of the ProX 500 and sPro series printers. The ProX 500 is the latest printer developed by 3D Systems with SLS technology. It is designed to increase the productivity and precision of the machine. It uses DuraForm® ProX™ materials to produce high-quality prototypes and parts in various areas. Furthermore, the sPro™ SLS series offers several different models to achieve different quality and print speed. Table 5.1 summarises the specifications of the ProX 500 and sPro™ 60 SD, sPro™ HD Base and sPro™ HD-HS. Table 5.2 summarises the specifications of the sPro™ 140 Base, sPro™ 140 HS, sPro™ 230 Base and sPro™ 230 HS. Fig. 5.1 shows the sPro™ 60 HD printer.

5.1.3 *The SLS® process*

The SLS® process creates 3D objects, layer-by-layer, from computer-aided design (CAD) data using powdered materials with heat generated by a CO_2 laser within the SLS machine. CAD data files in the STL file format are first transferred to the SLS machine systems where they are sliced. From this point, the SLS® process (see Fig. 5.2) begins and operates as follows:

(1) A thin layer of heat-fusible powder is deposited onto the part-building chamber.
(2) The bottom-most cross-sectional slice of the CAD part to be fabricated is selectively "drawn" (or scanned) on the layer of powder by a CO_2 laser. The interaction of the laser beam with the powder elevates the temperature to glass-transition temperature, fusing the powder particles to form a solid mass. The intensity of the laser beam is modulated to sinter the powder only in areas defined by the part's geometry. Surrounding powder remains a loose compact and serves as natural supports.
(3) When the cross section is completely "drawn", an additional layer of powder is deposited via a roller mechanism on top of the previously scanned layer. This prepares the next layer for scanning.

(4) Steps 2 and 3 are repeated, with each layer fusing to the layer below it. Successive layers of powder are deposited and the process is repeated until the part is complete.

Figure 5.1: sPro™ 60 HD printer.
Courtesy of 3D Systems

Table 5.1: Specifications of the ProX 500 and sPro™ 60 SD, sPro™ HD Base and sPro™ HD-HS.

Models	ProX 500	sPro™ 60 SD	sPro™ 60 HD Base	sPro™ 60 HD-HS
Build envelope capacity (XYZ)	15 × 13 × 18 in. (381 × 330 × 457 mm)	15 × 13 × 18 in. (381 × 330 × 437 mm), 15.2 US gal (57.5l)		
Powder layout	Variable Speed Counter Rotating Roller	Precision Counter Rotating Roller		
Layer thickness range (typical)	0.003–0.006 in. (0.08–0.15 mm) (0.004 in., 0.10 mm)	0.003 in. (Min 0.08 mm); (Max 0.006 in.) (0.15 mm), (0.004 in., 0.10 mm)		Min 0.003 in. (0.08 mm); Max 0.06 in. (0.15 mm), (0.0047 in.; 0.1 and 0.12 mm)
Imaging system	ProScan DX Digital High Speed	High Torque Scanning Motors (analogue)	ProScan™ CX (digital)	ProScan™ DX Dual Mode High Speed (digital)

Courtesy of 3D Systems

Table 5.1: (*Continued*) Specifications of the ProX 500 and sPro™ 60 SD, sPro™ HD Base and sPro™ HD-HS.

Models	ProX 500	sPro™ 60 SD	sPro™ 60 HD Base	sPro™ 60 HD-HS
Scanning speed	Fill, 500 in./s (12.7 m/s) Outline, 200 in./s (5 m/s)	240 in./s (6 m/s)	200 in./s (5 m/s)	240 and 480 in./s (6 and 12 m/s)
Laser power/ type	100 W/CO_2	30 W/CO_2	30 W/CO_2	70 W/CO_2
Volume build rate	2 L/h	0.9 L/h	1.0 L/h (60 cu in./h)	1.8 L/h (110 cu in./h)
Electrical requirements	208 VAC/ 7.5 kVA, 50/60 Hz, 1PH	240 V/12.5 kVA, 50/60 Hz AC 50/60 Hz, 3-phase (System)		
System warranty	–	1-year warranty, under 3D Systems' purchase terms and conditions		

Table 5.2: Specifications of the sPro 140 Base, sPro 140 HS, sPro 230 Base and sPro 230 HS.

Models	sPro™ 140 Base	sPro™ 140 HS	sPro™ 230 Base	sPro™ 230 HS
Build envelope capacity (XYZ)	22 × 22 × 18 in. (550 × 550 × 460 mm), 8500 cu in (139l)		22 × 22 × 30 in. (550 × 550 × 750 mm), 13,900 cu in (227l)	
Powder layout	Precision Counter Rotating Roller			
Layer thickness range (typical)	0.003 in. (Min 0.08 mm); Max 0.006 in. (0.15 mm), (0.004 in., 0.10 mm)			
Imaging system	ProScan™ Standard Digital Imaging System	ProScan™ GX Dual Mode High Speed Digital Imaging System	ProScan™ Standard Digital Imaging System	ProScan™ GX Dual Mode High Speed Digital Imaging System
Scanning speed	400 in./s (10 m/s)	400 in./s (15 m/s) (600 and 400 in./s)	400 in./s (10 m/s)	400 in./s (15 m/s) (600 and 400 in./s)
Laser power/ type	70 W/CO_2	200 W/CO_2	70 W/CO_2	200 W/CO_2
Volume build rate	3.0 L/h (185 cu in./h)	5.0 L/h (300 cu in./h)	3.0 L/h (185 cu in./h)	5.0 L/h (300 cu in./h)
Electrical requirements	208 V/17 kVA, 50/60 Hz AC 50/60 Hz, 3-phase (System)			
System warranty	1-year warranty, under 3D Systems' purchase terms and conditions			

Courtesy of 3D Systems

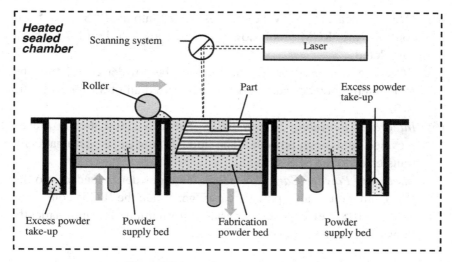

Figure 5.2: Schematic of the SLS process.

As SLS® materials are in powdered form, the powder that is not melted or fused during processing serves as a customised, inherent built-in support structure. Thus, there is no need to create additional support structures within the CAD design and, therefore, no post-build removal of these supports is required. After the SLS® process, the part is removed from the build chamber and the loose powder simply falls away. SLS® parts may then require some post-processing or secondary finishing, such as sanding, lacquering and painting, depending on the application of the prototype built.

The sPro SLS® system contains these hardware components[1]:

(1) *Sinterstation Pro SLS System.* Manufactures part(s) from 3D CAD data.
(2) *Rapid Change Module (RCM).* Build module mounted on wheels for quick and easy transfer between the Sinterstation, the Offline Thermal Station (OTS) and the Break Out Station (BOS).
(3) *Nitrogen Generator.* Delivers a continuous supply of nitrogen to the SLS system to keep the fabrication inert and prevent oxidation.

(4) *OTS*. Pre-heats the RCM before it is loaded into the SLS system and controls the RCM cool down process after a build has been completed.
(5) *BOS*. The built parts are extracted from the powder cake here. The non-sintered powder automatically gets sifted and transferred to the IRS.
(6) *Integrated Recycling Station (IRS)*. The IRS automatically blends recycled and new powder. The mixed powder is automatically transferred to the SLS system.
(7) *Intelligent Powder Cartridge (IPC)*. New powder is loaded into the IRS from a returnable powder cartridge. When the IPC is connected to the IRS, electronic material information is automatically transferred to the SLS system.

The software and system controller for Sinterstation® HiQ™ Series SLS® System includes the proprietary SLS system software running on Microsoft's Windows XP operating system. The software that comes with Sinterstation® Pro SLS® system includes the following:

- Build Setup and Sinter (included).
- SinterscanTM (optional) software provides more uniform properties in x- and y-directions and improved surface finish.
- RealMonitorTM (optional) software provides advanced monitoring and tracking capabilities.

5.1.4 *Materials*

The materials used in SLS® system can be broadly classified into two groups: DuraForm® materials and CastForm™ materials.[2]

The DuraForm® group consists of the following materials: DuraForm® GF material, DuraForm® PA material, DuraForm® EX material, DuraForm® Flex plastic, DuraForm® FR 100 material, DuraForm® HST Composite material, DuraForm® ProX™ material and DuraForm® AF plastic. These are discussed in the following:

(1) *DuraForm*® *GF plastic.* These are glass-filled polyamide (nylon) material for tough real-world physical testing and functional applications. The features of the material are as follows: excellent mechanical stiffness, elevated temperature resistance, dimensional stability, easy-to-process and relatively good surface finish. Applications for the material include housings and enclosures, consumer sporting goods, low-to-medium batch size manufacturing, functional prototypes, parts requiring stiffness and thermally stressed parts.

(2) *DuraForm*® *PA plastic.* These are durable polyamide (nylon) material for general physical testing and functional applications. The features of the material are as follows: excellent surface resolution and feature detail, easy-to-process, compliant with USP Class VI testing, compatible with autoclave sterilisation, and good chemical resistance and low moisture absorption. Applications for the material include producing complex, thin wall ductwork, for example motorsports, aerospace, impellers and connectors, consumer sporting goods, vehicle dashboards and grilles, snap-fit designs, functional prototypes that approach end-use performance properties, medical applications requiring USP Class VI compliance, parts requiring machining or joining with adhesives.

(3) *DuraForm*® *EX plastic.* These are impact-resistant plastic offering the toughness of injection-moulded thermoplastics and are suitable for rapid manufacturing. They are available in either natural (white) or black colours. Features of the material are that it offers the toughness and impact resistance of injection-moulded ABS and polypropylene. Applications for the material include complex, thin-walled ductwork, motorsports, aerospace and unmanned air vehicles (UAVs), snap-fit designs, hinges, vehicle dashboards, grilles and bumpers.

(4) *DuraForm*® *Flex plastic.* This is a thermoplastic elastomer material with rubber-like flexibility and functionality. Features of the material are as follows: flexible, durable with good tear resistance, variability of Shore A hardness using the same material, good powder recycle characteristics, good surface finish and feature

detail. The applications for the material include athletic footwear and equipment, gaskets, hoses and seals, simulated thermoplastic elastomer, cast urethane, silicone and rubber parts.

(5) *DuraForm® AF plastic.* These are polyamide (nylon) material with metallic appearance for real-world physical testing and functional use. Features of the material are as follows: metallic appearance with nice surface finish, good powder recycle characteristics, excellent mechanical stiffness, easy-to-process and dimensional stability. Applications for the material include housings and enclosures, consumer products, thermally stressed parts and plastic parts requiring a metallic appearance.

(6) *DuraForm® FR 100 material.* This is a halogen and antimony-free, flame retardant engineering plastic. It is suitable for AM of aerospace parts and parts requiring UL 94-V-0 compliance.

(7) *DuraForm® HST Composite material.* This is a fibre-reinforced engineering plastic with high stiffness, strength and temperature resistance.

(8) *DuraForm® ProX™ material.* This is an extra-strong engineered production plastic. It is used to produce durable functional prototypes with superior mechanical properties. It was developed in tandem with ProX 500 printer to print smoother wall surfaces comparable to injection-moulded part.

Lastly, the CastForm™ PS material. This material directly produces complex investment casting patterns without tooling. Features of this "foundry friendly" material include: functions such as foundry wax, low residual ash content (<0.02%), short burnout cycle, easy-to-process plastic and good plastic powder recycle characteristics. Applications of the material include creating complex investment casting patterns, indirectly producing reactive metals (such as titanium and magnesium), near net-shaped components, low-melting point metals (such as aluminium, magnesium and zinc), ferrous and non-ferrous metals. Smaller parts can be joined to create very large patterns, which are sacrificial and expendable.

5.1.5 *Principle*

The SLS® process is based on the following two principles:

(1) Parts are built by sintering when a CO_2 laser beam hit a thin layer of powdered material. The interaction of the laser beam with the powder raises the temperature of the powder prior to its melting point, resulting in particle bonding, fusing the particles to themselves and the previous layer to form a solid. This is the basic principle of sinter bonding.

(2) The building of the part is done layer-by-layer. Each layer of the building process contains the cross sections of one or many parts. The next layer is then built directly on top of the sintered layer after an additional layer of powder is deposited via a roller mechanism.

The packing density of particles during sintering affects the part density. In studies of particle packing with uniform-sized particles[3] and particles used in commercial sinter bonding,[4] packing densities are found to range typically from 50% to 62%. Generally, the higher the packing density, the better the mechanical properties can be expected. However, it must be noted that scan pattern and exposure parameters are also major factors in determining the part's mechanical properties.

In the process of sinter bonding, particles in each successive layer are fused to each other and to the previous layer by raising their temperature with the laser beam to above the glass-transition temperature. The glass-transition temperature is the temperature at which the material begins to soften from a solid state to a jelly-like condition. This often occurs just prior to the melting temperature at which the material will be in a molten or liquid state. As a result, the particles begin to soften and deform owing to its weight and cause the surfaces in contact with other particles or solid to deform and fuse together at these contact surfaces. One major advantage of sintering over melting and fusing is that it joins powder particles into a solid part without going into the liquid phase, thus avoiding the distortions caused by the flow of molten material during fusion. After cooling, the powder particles are connected in a matrix that has approximately the density of the particle material.

As the sintering process requires the machine to bring the temperature of the particles to the glass-transition temperature, the energy required is considerably high. The energy required to sinter bond a similar layer thickness of material is approximately between 300 and 500 times higher than that required for photopolymerisation.[5,6] This high-power requirement can be reduced by using auxiliary heaters to raise the powder temperature to just below the sintering temperature during the sintering process. However, an inert gas environment is needed to prevent oxidation or explosion of the fine powder particles. Cooling is also necessary for the chamber gas.

The parameters which affect the performance and functionalities are the properties of the powdered materials and its mechanical properties after sintering, the accuracy of the laser beam, the scanning pattern, the exposure parameters and the resolution of the machine.

5.1.6 *Strengths and weaknesses*

The strengths of the SLS® process include:

(1) *Good part stability.* Parts are created within a precise controlled environment. The process and materials provide for functional parts to be built directly.
(2) *Wide range of processing materials.* In general, any material in powder form can be sintered on the SLS. A wide range of materials including nylon, polycarbonates, metals and ceramics are available directly from 3D Systems, thus providing flexibility and a wide scope of functional applications.
(3) *No part supports required.* The system does not require CAD-developed support structures. This saves the time required for support structure building and removal.

(4) *Little post-processing required.* The finish of the part is reasonably fine and requires only minimal post-processing such as particle blasting and sanding.

(5) *No post-curing required.* The completed laser sintered part is generally solid enough and does not require further curing.

(6) *Advanced software support.* The new version 2.0 software uses a Windows NT-style graphical user interface (GUI). Apart from the basic features, it allows for streamlined parts scaling, advanced non-linear parts scaling, in-progress part changes and build report utilities.[7] It is available in different foreign languages.

The weaknesses of the SLS® process include:

(1) *Large physical size of the unit.* The system requires a relatively large space to house it. Apart from this, additional storage space is required to house the inert gas tanks that are required for each build.

(2) *High power consumption.* The system requires high power consumption due to the high wattage of the laser required to sinter the powder particles together.

(3) *Poor surface finish.* The as-produced parts tend to have poorer surface finish due to the relatively large particle sizes of the powders used.

5.1.7 *Applications*

The SLS® system can produce a wide range of parts in a broad variety of applications, including the following:

(1) *Concept models.* Physical representations of designs used to review design ideas, form and style.

(2) *Functional models and working prototypes.* Parts that can withstand limited functional testing, or fit and operate within an assembly.

(3) *Polycarbonate (RapidCasting™) patterns.* Patterns produced using polycarbonate, and then cast in the metal of choice through the standard investment casting process. These build faster than wax patterns and are ideally suited for designs with thin walls and fine features. These patterns are also durable and heat resistant.

(4) *Metal tools (RapidTool™).* Direct rapid prototype of tools of moulds for small or short production runs.

(5) *Aerospace ducting.* With parts and components with high precision and high strength produced by SLS® in ducting system, AM's products are already been used in many different aircrafts.

5.1.8　Research and development

Primary research continues to focus on new and advanced materials while further improving and refining SLS® process, software and system. Currently, ProX 500 is one of the latest SLS® machines while 3D Systems is still working to improve the build volume as well as to reduce the cost.

5.2　SLM Solutions' Selective Laser Melting

5.2.1　Company

SLM Solutions GmbH, formerly known as MTT Technologies GmbH, Lübeck, focuses on the development of their selective laser melting (SLM) systems. It started as one of the most popular processes — vacuum casting which the company introduced as the very first supplier in 1987. The SLM systems was first introduced to the market in 2000 and it built tools and single individual engineering and implant parts in pure, dense steel, cobalt-chrome and titanium from CAD data files. The address of SLM Solutions GmbH head office is Roggenhorster Strasse 9c, D-23556 Lübeck, Germany.

5.2.2　Products

Based on the SLM technology, SLM Solutions has developed a series of 3D printers that can process most metals. They are SLM® 125HL (see Fig. 5.3), SLM® 280HL (see Fig. 5.4) and SLM® 500HL (see Fig. 5.5). These machines make it possible to produce 100% dense metal parts from customary metal powder. The parts or specialty tools are built layer-by-layer (30-μm thickness). Metal powder (e.g., stainless steel 1.4404) is locally melted by an intensive infrared laser beam that traces the layer geometry. These machine specifications are listed in Table 5.3. Besides, SLM Solutions also features its automatic powder sieving station known as the SLM® PSA500 (see Fig. 5.5).

Figure 5.3: SLM® 125HL printer.
Courtesy of SLM Solutions

Figure 5.4: SLM® 280HL printer.
Courtesy of SLM Solutions

Figure 5.5: SLM® PSA500 automatic powder sieving station (left) and SLM® 500HL printer (right).
Courtesy of SLM Solutions

Table 5.3: Specifications for SLM® 125HL, SLM® 280HL and SLM® 500HL.

Models	SLM® 125HL	SLM® 280HL	SLM® 500HL
Build Envelope (L × W × H)	125 × 125 × 125 mm³ reduced by substrate plate thickness	280 × 280 × 365 mm³ reduced by substrate plate thickness	500 × 280 × 365 mm³ reduced by substrate plate thickness
3D Optics Configuration (IPG fibre laser)	Single (1 × 400 W)	Single (1 × 400 W), Twin (2 × 400 W), Dual (1 × 400 W and 1 × 1000 W); Single (1 × 700 W), Twin (2 × 700 W), Dual (1 × 700 W and 1 × 1000 W)	Twin (2 × 400 W), Quad (4 × 400 W), Twin (2 × 700 W), Quad (4 × 700 W)
Build Rate	up to 25 cm³/h	up to 55 cm³/h	up to 105 cm³/h
Variable Layer Thickness	20–75 µm, 1-µm increments	20–75 µm	20–75 µm
Min. Feature Size	140 µm	150 µm	150 µm
Beam Focus Diameter	70–100 µm	80–115 µm	80–115 µm
Max. Scan Speed	10 m/s	10 m/s	10 m/s
Average Inert Gas Consumption in Process (argon)	2 L/min	2.5 L/min	5–7 L/min
Average Inert Gas Consumption Purging (argon)	70 L/min	70 L/min	70 L/min
Compressed Air Requirement/ Consumption	ISO 8573-1:2010 [1:4:1], 50 L/min @ 6 bar	ISO 8573-1:2010 [1:4:1], 50 L/min @ 6 bar	ISO 8573-1:2010 [1:4:1], 50 L/min @ 6 bar
Dimensions (L × W × H)	1400 × 900 × 2460 mm	3050 × 1050 × 2850 mm(incl. PSH100)	5200 × 2800 × 2700 mm(incl. PSX, PRS)
Weight (incl. without powder)	Approx. 750/700 kg	Approx. 1500/1300 kg	Approx. 3100/2400 kg
E-Connection/ Power Input	400 V 3NPE, 32 A, 50/60 Hz, 3 kW	400 V 3NPE, 32 A, 50/60 Hz, 3.5–5.5 kW	400 V 3NPE, 64 A, 50/60 Hz, 8–10 kW

5.2.3 *Process*

The SLM process adopts many of the conventions of the layer-based manufacturing processes. What differs is that the system can be used on a wide range of powdered metal materials. A component is split into layers and each of those layers is built on top of each other and fused together, layer after layer, until the part's form is built.

The SLM process uses a laser beam, controlled using optic lenses to scan a laser spot across the surface of a layer of powdered metal to build each layer. The metal powder is melted rather than just simply sintered together, thus giving parts which are almost 100% dense (or solid). The resulting part has much greater strength and dimensional accuracy than parts that are built by laser sintering.[8,9] The schematic diagram in Fig. 5.6 illustrates the SLM process.

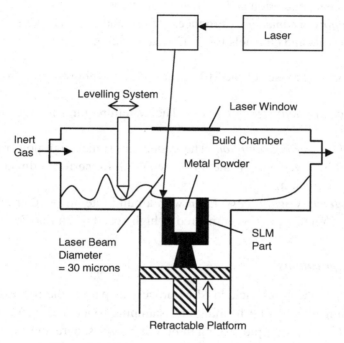

Figure 5.6: Schematic diagram of the SLM process.

5.2.4 *Strengths and weaknesses*

The main strengths of the SLM Solutions' SLM are as follows:

(1) *High-quality metal parts.* Builds high-quality parts and tooling inserts from metal powders. Allows homogeneous build-up of components and tool cavities up to 100% density depending on requirements.

(2) *Large range of metal materials.* The system can be used to build almost any type of metal: stainless steel, tool steel, titanium, aluminium, cobalt-chrome, copper, platinum and various non-ferrous metals. The system can also be used to shape gold.

(3) *Relatively lesser processes and low cost.* This is because no post-processing like infiltration is required.

(4) *High accuracy.* High-resolution process, dimensionally accurate, low heat generation and no distortion of parts.

(5) *Complex geometries.* Produces tools and inserts with internal undercuts and channels for conformal cooling.

The major weaknesses of the SLM process are as follows:

(1) *Large physical size of the unit.* The system requires a large space to house.

(2) *High power consumption.* The system consumes high power due to the high wattage of the laser required to perform direct metal powder melting.

(3) *Relatively slow process.* Even with a build rate up to 70 cm^3/h, it is still a lengthy process compared to high-speed machining.[8]

5.2.5 *Applications*

Industrial applications include SLM process as part of the process chain for building sheet metal forming and stamping tools. The SLM process can also be used for rapid tooling in building inserts, core and cavities for plastic injection moulding applications. Other applications include building functional prototypes and parts with specialised or dedicated metal, for example aluminium alloy, for the aerospace and biomedical industries.

5.2.6 *Research and development*

SLM Solutions always specifically addresses the latest developments in SLM technology. It aims to keep improving the capabilities and coming up with new breakthroughs for SLM.

5.3   3D Systems' CJP Technology

5.3.1   *Company*

Originally invented, patented and developed at the Massachusetts Institute of Technology (MIT) in 1993, 3D Printing technology (3DP™) forms the basis of Z Corporation's licensed prototyping process. Z Corp. pioneered the commercial use of 3DP technology, developing 3D printers that leading manufacturers used to produce early concept models and product prototypes for a broad range of applications.

Z Corporation was incorporated in 1994 by Hatsopoulos, Walter Bornhost, Tim Anderson and Jim Brett. It commercialised its first 3D printer, the Z™402 System, based on the 3DP technology in 1997. In 2000, Z Corp. launched its first colour 3D printer and subsequently introduced high-definition 3DP (HD3DP) in 2005. It was acquired by 3D Systems in April 2013 and 3D Systems' ProJet x60 series replaced the ZPrinter from Z Corp. The technology was also renamed as CJP. 3D Systems' details are found in Section 3.1.1.

5.3.2   *Products*

3D Systems' current products are ProJet® CJP 260*C*, ProJet® CJP 360, ProJet® CJP 460*Plus*, ProJet® CJP 660*Pro* and ProJet® CJP 860*Pro* (see Fig. 5.7). The ProJet® CJP 260*C* is the upgraded version from the previous 160 series with the addition of basic 3-channel CMY colour 3DP. Meanwhile, ProJet® CJP 360 is the enhancement of the 160 series and the 260*C* series. It expands the build volume to 203 × 254 × 203 mm and reduces the cost of printing. Hence, it is often used to produce architectural modelling and medium-sized prototypes. Moreover, ProJet® CJP 460*Plus* further improves the machines with safer build materials, active dust control and zero liquid waste. Furthermore, ProJet® CJP 660*Pro* has a larger building volume of 254 × 381 × 203 mm, incorporating 3D System's 4-channel CMYK full colour 3DP to print high-resolution prototypes. It is mainly used by designers and researchers

Figure 5.7: ProJet® 860*Pro*.
Courtesy of 3D Systems

Table 5.4: Specifications of ProJet® CJP 260*C*, ProJet® CJP 360, ProJet® CJP 460*Plus*, ProJet® CJP 660*Pro* and ProJet® CJP 860*Pro*.

Models	ProJet® CJP 260*C*	ProJet® CJP 360	ProJet® CJP 460*Plus*	ProJet® CJP 660*Pro*	ProJet® CJP 860*Pro*
Resolution	300 × 450 dpi	300 × 450 dpi	300 × 450 dpi	600 × 540 dpi	600 × 540 dpi
Colour	CMY	White (mono-chrome)	CMY	Full CMYK	Full CMYK
Pastel or vibrant colour options	No	No	No	Yes	Yes
Minimum feature size	0.04 in. (1 mm)	0.006 in. (0.15 mm)	0.006 in. (0.15 mm)	0.004 in. (0.1 mm)	0.004 in. (0.1 mm)
Layer thickness	0.004 in. (0.1 mm)	0.004 in. (0.1 mm)	0.004 in. (0.1 mm)	0.004 in. (0.1 mm)	0.004 in. (0.1 mm)
Vertical build speed	0.8 in./h (20 mm/h)	0.8 in./h (20 mm/h)	0.9 in./h (23 mm/h)	1.1 in./h (28 mm/h)	0.2–0.6 in./h (5–15 mm/h); Speed increases with volume of prototypes
Prototypes per build	10	18	18	36	96
Draft printing mode (Monochrome)	No	No	No	Yes	Yes
Net build volume (XYZ)	9.3 × 7.3 × 5 in. (236 × 185 × 127 mm)	8 × 10 × 8 in. (203 × 254 × 203 mm)	8 × 10 × 8 in. (203 × 254 × 203 mm)	10 × 15 × 8 in. (254 × 381 × 203 mm)	20 × 15 × in. (508 × 381 × 229 mm)

Table 5.4: (*Continued*) Specifications of ProJet® CJP 260*C*, ProJet® CJP 360, ProJet® CJP 460*Plus*, ProJet® CJP 660*Pro* and ProJet® CJP 860*Pro*.

Models	ProJet® CJP 260C	ProJet® CJP 360	ProJet® CJP 460Plus	ProJet® CJP 660Pro	ProJet® CJP 860Pro
Build materials	VisiJet® PXL™	VisiJet® PXL™	VisiJet® PXL™	VisiJet® PXL™	VisiJet® PXL™
Number of jets	604	304	604	1520	1520
Number of print heads	2	1	2	5	5
Automated setup and self-monitoring	Yes	Yes	Yes	Yes	Yes
Material recycling	Yes	Yes	Yes	Yes	Yes
Automatic build platform clearing	No	No	Yes	Yes	Yes
Part cleaning	Accessory	Integrated	Integrated	Integrated	Accessory
Integrated materials	Yes	Yes	Yes	Yes	Yes
Intuitive control panel	Yes	Yes	Yes	Yes	Yes
Input data file formats supported	STL, VRML, PLY, 3DS, FBX, ZPR				
Client operating system	Windows® 7 and Vista®				
Operating temperature range	55°F–75°F (13°C–24°C)	55°F–75°F (13°C–24°C)	55°F–75°F (13°C–24°C)	55°F–75°F (13°C–24°C)	55°F–75°F (13°C–24°C)
Operating humidity range	20%–55% (non-cond.)	20%–50% (non-cond.)	20%–50% (non-cond.)	20%–50% (non-cond.)	20%–50% (non-cond.)
Printer dimensions	29 × 31 × 55 in. (74 × 79 × 140 cm)	48 × 31 × 55 in. (122 × 79 × 140 cm)	48 × 31 × 55 in. (122 × 79 × 140 cm)	74 × 29 × 57 in. (188 × 74 × 145 cm)	47 × 46 × 68 in. (119 × 116 × 162 cm)
Printer weight	365 lbs (165 kg)	395 lbs (179 kg)	425 lbs (193 kg)	750 lbs (340 kg)	800 lbs (363 kg)
Electrical	90–100 V, 7.5 A 110–120 V, 5.5 A 208–240 V, 4.0 A	90–100 V, 7.5 A 110–120 V, 5.5 A 208–240 V, 4.0 A	90–100 V, 7.5 A 110–120 V, 5.5 A 208–240 V, 4.0 A	100–240 V, 15–7.5 A	100–240 V, 15–7.5 A
Noise Building Core Recovery Vacuum (open) Fine Decoring	57 dB 66 dB 86 dB –	57 dB 66 dB 86 dB 80 dB	57 dB 66 dB 86 dB 80 dB	57 dB 66 dB 86 dB 80 dB	57 dB 66 dB 86 dB –
Certifications	CE, CSA	CE, CSA	CE, CSA	CE, CSA	CE, CSA

to print professional models, art pieces and more. In the meantime, ProJet® CJP 860*Pro* is targeted to produce larger models. Table 5.4 summarises the specifications of ProJet® CJP 260*C*, ProJet® CJP 360, ProJet® CJP 460*Plus*, ProJet® CJP 660*Pro* and ProJet® CJP 860*Pro*.

5.3.3 *Process*

3D Systems' CJP technology, formerly known as the 3DP technology, creates 3D physical prototypes by solidifying layers of deposited powder using a liquid binder. The CJP process is shown in Fig. 5.8.

(1) The machine spreads a layer of powder from the feed box to cover the surface of the build piston. The printer then prints binder solution onto the loose powder, forming the first cross section. For multi-coloured parts, each of the four print heads deposits a different colour binder, mixing the four colour binders to produce a spectrum of colours that can be applied to different regions of a part.

(2) The powder is glued together by the binder at where it is printed. The remaining powder remains loose and supports the following layers that are spread and printed above it.

(3) When the cross section is completed, the build piston is lowered, a new layer of powder is spread over its surface and the process is repeated. The part grows layer-by-layer in the build piston until the part is complete, completely surrounded and covered by loose powder. Finally, the build piston is raised and the loose powder is vacuumed away, revealing the complete part.

(4) Once a build is completed, the excess powder is vacuumed away and the parts are lifted from the bed. Once removed, parts can be finished in a variety of ways to suit your needs. For a quick design review, parts can be left raw or "green". To quickly produce a more robust model, parts can be dipped in wax. For a robust model that can be sanded, finished and painted, the part can be infiltrated with a resin or urethane.

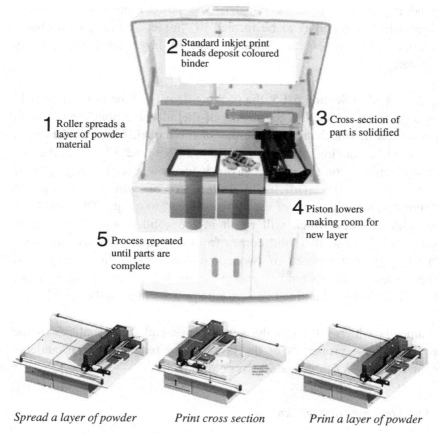

2 Standard inkjet print heads deposit coloured binder

1 Roller spreads a layer of powder material

3 Cross-section of part is solidified

4 Piston lowers making room for new layer

5 Process repeated until parts are complete

Spread a layer of powder   *Print cross section*   *Print a layer of powder*

Figure 5.8: Illustration of CJP.
Courtesy of 3D Systems

5.3.4 *Materials*

With 3D Systems' CJP, VisiJet® PXL™ material is used by ProJet® x60 series 3D printers to produce strong, high performance and full colour models and prototypes.

5.3.5 *Principle*

3D Systems' CJP creates parts by a layered printing process and adhesive bonding, based on sliced cross-sectional data. A layer is created by adding another layer of powder. The powder layer is selectively joined where the part is to be formed by "inkjet" printing of a binder material. The process is repeated layer-by-layer until the part is complete.

As described in Section 5.1.5, the packing density of the powder particle has a profound impact on the results of the adhesive bonding which in turn affects the mechanical properties of the model. Like powders used on the SLS, packing densities are from 50% to 62%.[4] When the ink droplet impinges on the powder layer, it forms a spherical aggregate of binder and powder particles. Capillary forces will cause adjacent aggregates, including that of the previous layer, to merge. This will form the solid network which will result in the solid model. The binding energy for forming the solid comes from the liquid adhesive droplets. This energy is composed of two components, its surface energy and kinetic energy. As this binding energy is low, it is about 10^4 times[4] more efficient than sinter binding in converting powder to a solid object.[3]

Parameters which influence the performance and functionalities of the process are the properties of the powder, the binder material, and the accuracy of the *XY* table and *Z*-axis control.

5.3.6 *Strengths and weaknesses*

The key strengths of the CJP systems are as follows:

(1) *High speed.* CJP are high-speed printers. Each layer is printed in seconds, reducing the prototyping time of a hand-held part to 1–2 hours.
(2) *Versatile.* Parts are currently used for automotive, packaging, education, footwear, medical, aerospace and telecommunications industries. Parts are used in every step of the design process for communication, design review and limited functional testing. Parts can be infiltrated if necessary, offering the opportunity to produce parts with a variety of material properties to serve a range of modelling requirements.

(3) *Simple to operate.* The office system is straightforward to operate and does not require a designated technician to build a part. The system is based on the standard, off the shelf components developed for the inkjet printer industry, resulting in a reliable and dependable 3D printer.

(4) *Minimal wastage of materials.* Powder that is not printed during the cycle can be reused.

(5) *Colour.* It enables complex colour schemes for AM parts from a full 24-bit palette of colours to be made possible.

The limitations of the CJP systems include:

(1) *Limited functional parts.* Relative to the SLS, parts built by CJP are much weaker, thereby limiting the functional testing capabilities.

(2) *Poor surface finish.* Parts built by CJP have relatively poorer surface finish and post-processing is frequently required.

5.3.7 *Applications*

The CJP process can be used in the following areas:

(1) *Concept and functional models.* Creating physical representations of designs used to review design ideas, form and style. With the infiltration of appropriate materials, it can also create parts that are used for functional testing, fit and performance evaluation.

(2) *CAD-Casting metal parts.* CAD-Casting is a term used to connote a casting process where the mould is fabricated directly from a computer model with no intermediate steps. In this method, a ceramic shell with integral cores may be fabricated directly from a computer model. This results in tremendous streamlining of the casting process.

(3) *Direct metal parts.* Metal parts in a range of material including stainless steel, tungsten and tungsten carbide can be created from metal powder with CJP process. Printed parts are post-processed using techniques borrowed from metal injection moulding.

(4) *Structural ceramics.* CJP can be used to prepare dense alumina parts by spreading sub-micron alumina powder and printing a latex binder. The green parts are then isostatically pressed and sintered to densify the component. The polymeric binder is then removed by thermal decomposition.

(5) *Functionally gradient materials.* CJP can create composite materials as well. For example a ceramic mould can be 3D printed, filled with particulate matter and then pressure infiltrated with a molten material. Silicon carbide reinforced aluminium alloys can be produced directly by 3DP a complex SiC substrate and infiltrating it with aluminium, allowing localised control of toughness.

5.3.8 *Research and development*

3D Systems is launching a consumer 3D printer that produces ceramic parts ready for firing and glazing. Using the CJP technology, the CeraJet is capable of producing intricate and detailed ceramic objects that many artists and designers would like to work with.

5.4 BeAM's LMD Systems

5.4.1 *Company*

"Be Additive Manufacturing" (BeAM) is a French startup founded in December 2012. The company pioneered as the first European manufacturer of AM machine based on the laser metal deposition (LMD) technology (initially known as the EasyCLAD Systems), in which the technology was developed by IREPA LASER, a regional research and development (R&D) and training centre that specialises particularly in laser processes and materials. With the LMD technology, BeAM is able to provide a qualified repair service, repairing previously irreparable plane engine parts. By repairing aircraft engine components, BeAM has reached the highest level of development for industrial applications in the field of AM. In the year 2015, BeAM was awarded the Aeronautics Award under the category of "Innovation". In the meantime, BeAM has considered to expand into new markets such as defense, energy, shipbuilding and railroad industries. BeAM has its head office located in Pôle API-Parc d'Innovation, 67400 Illkirch-Graffenstaden, France.

5.4.2 *Products*

BeAM offers two ranges of its LMD-based AM machines — the Mobile (Fig. 5.9) and the Magic 2.0 (Fig 5.10). Suitable for small and medium volume, the Mobile is optimised for direct manufacturing and repairing of thin and complex parts. Meanwhile, the Magic 2.0, released in year 2015, is a derivative of the Magic 1.0, having improved maintainability, accessibility, productivity and bigger area in 3 axes up to 1.4-m high. The Magic 2.0 was developed for the state of the art industries that require specific working areas for direct AM to repair large metallic parts. Complemented with the BeAM software, Magic 2.0 is specifically designed for manufacturing and repairing complex parts in controlled atmosphere (<40-ppm O_2 and <50-ppm H_2O). Both Mobile and Magic 2.0 systems are offered in continuous 3-axis configurations, with continuous 5-axis configurations as an option. Additionally, it would be good to note that the MAGIC 2.0 machine is much larger in size as compared to the Mobile machine. The specifications of Mobile and Magic 2.0 are summarised in Table 5.5.

<table>
<tr><td>Figure 5.9: Mobile.
Courtesy of BeAM French Company</td><td>Figure 5.10: Magic 2.0.
Courtesy of BeAM French Company</td></tr>
</table>

Table 5.5: Specifications of Mobile and Magic 2.0.

Models	Mobile	Magic 2.0
Technical Characteristics		
Machine dimensions	$L = 1390 \times 1 = 1270 \times H = 2400$ mm	$L = 3600 \times 1 = 2000 \times H = 3100$ mm
Weight	–	8 tonnes (approximate)
Axis	Continuous 5-axis ($X\ Y\ Z + B\ C$)	Continuous 5-axis ($X\ Y\ Z + B\ C$)
Working area (X, Y, Z) mm	$400 \times 250 \times 200$	$1200 \times 800 \times 800$ (3-axis option: $500 \times 700 \times 1400$)
Power supply	400 V/63 A Three-phases	400 V (10%)/50 Hz Three-phases
Numerical control (NC)	Power Automation	SIEMENS 840D
Dialog protocol	FTP/Ethernet/TCP-IP	FTP/Ethernet/TCP-IP
Number of deposition nozzles	1	1–2
Powder feeder	1–5 bowls (capacity: 1.5 L each)	1–5 bowls (capacity: 1.5 L each)
Controlled atmosphere enclosure (optional)	2.5 m^3, equipped with recycling and purification system	10 m^3, equipped with recycling and purification system
Controlled atmosphere	Less than 40-ppm O_2, less than 50-ppm H_2O	Less than 40-ppm O_2, less than 50-ppm H_2O
Options	Renishaw MP250 probing tool, Process monitoring TPSH	Elevating system, Renishaw MP250 probing tool, Process monitoring TPSH, Automatic or semi-automatic loading/unloading system
Technology		
Laser	Fibre Laser IPG 500 W	Laser (fibre) IPG 2 KW/3 KW/4 KW 2 outputs
LMD deposition system	10 Vx	10 Vx or 24 Vx
Deposition volume, build rate	10–70 cm^3/h	10–300 cm^3/h
Deposit/layer thickness	0.1 mm/layer	0.1–0.8 mm/layer
Deposit width	0.8–1.2 mm	0.8 mm to more than 5 mm
Powder size	45–90 µm	45–90 µm

Table 5.5: (*Continued*) Specifications of Mobile and Magic 2.0.

Models	Mobile	Magic 2.0
BeAM CAD/CAM software		
Operating system	Windows 7/Windows 8/Windows 10	Windows 7/Windows 8/Windows 10
Network connection	Ethernet	Ethernet
Required applications	DELCAM PowerSHAPE, PowerMILL	DELCAM PowerSHAPE, PowerMILL
File types	STEP, IGES, ProE, Catia, SolidWorks, . . .	STEP, IGES, ProE, Catia, SolidWorks, . . .
Certifications		
Machine	CE	CE
Laser	EN 60825-1: Class 1	EN 60825-1: Class 1

5.4.3 *Process*

The BeAM's LMD technology allows direct manufacturing of metal parts directly from CAD files. The LMD process is conducted under a controlled atmosphere that has the oxygen and moisture levels stay below 40 and 50 ppm, respectively. Together with the usage of noble gases, the controlled environment keeps the part clean and prevents oxidation during the build process. Through the CLAD® nozzle, metal powders are injected and melted with a laser beam to generate a homogenous metallic layer (see Fig. 5.11). The LMD process comprises of the following steps:

(1) Tool trajectories are generated entirely by BeAM's CAD/CAM software, which was developed specially in conjunction with LMD technology. The software can also be used to develop manufacturing strategies, write .iso programmes as well as simulate construction or repair processes. The operator may select manufacturing strategies catered for the application from standard numerical files (STEP, IGES, etc.) or source files (ProE, CATIA, SolidWorks, etc.).

(2) Metal powders are injected and melted through a dedicated nozzle by a laser beam. Following the tool trajectories, successive layers of metal are deposited, manufacturing the part directly with excellent control of its dimensions and its material quality.

(3) When a layer is completed, the deposition nozzle and the platform move to continue with the next layer. The process is repeated layer-by-layer until the part is completed. Generally, the prototypes need additional finishing, but are fully dense products with excellent metallurgical properties.

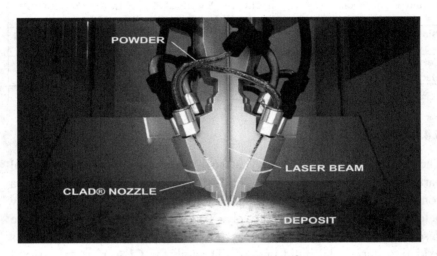

Figure 5.11: LMD Process with CLAD® nozzle.
Courtesy of BeAM French Company

5.4.4 *Materials*

Associating with BeAM's own software, BeAM machines can build multi-material structures with different benefits such as wear and abrasion resistance, thermal dissipation properties and so on. The BeAM machines generally have wide varieties of metallic material choices with the possibilities of other new materials.

5.4.5 *Principles*

LMD belongs to the powder directed energy deposition class in the AM technology and the process is also known as the direct metal laser deposition (DMLD), laser powder deposition (LPD), direct laser deposition (DLD), direct metal deposition (DMD), laser engineered net shaping (LENS), laser cladding, laser deposition welding (LDW) and powder fusion welding. The LMD process is based on the following two principles:

(1) A high-powered fibre laser focused onto a metal substrate, creating a molten puddle on the substrate surface. Powder is continuously injected into the molten puddle to increase material volume.
(2) A "printing" motion system moves a platform as the laser beam traces the cross section of the part being produced. After formation of a layer of the part, the platform and the machine's powder delivery nozzle adjust themselves prior to building next layer.

5.4.6 *Strengths and weaknesses*

The key strengths of BeAM's LMD systems are as follows:

(1) *Superior material properties.* The BeAM's LMD process is capable of producing fully dense metal parts. The resulting direct manufactured part has high metallurgical quality which is at least equal to foundry's metallurgical quality.
(2) *Complex parts.* Functional metal parts with complex features can be produced on the BeAM's LMD system.
(3) *Add material onto existing parts.* BeAM's LMD systems can not only fully build 3D metal parts, but also can add materials to existing parts. For example, adding of shapes and functions onto an existing part, or to repair worn/damaged component, or to combine BeAM's LMD with conventional manufacturing methods to create unique hybrid manufacturing solutions.
(4) *Blend materials.* BeAM's LMD systems can blend multiple powders together during the build process to construct parts with functional gradients or new alloys tailored to meet specific requirements.
(5) *Low powder cost.* The BeAM's LMD process can use a wide powder size range and thus powder costs are minimised, achieving economy of material.

(6) *Large working area.* LMD has no volume restriction in terms of working area as BeAM's LMD technology uses a dedicated nozzle to inject the metal powder instead of utilising a powder bed.

The primary limitations of the BeAM's LMD systems are as follows:

(1) *Geometrical complexity.* The BeAM's LMD process can manufacture relatively complicated shapes, but is limited in complexity compared to casting or powder bed processes.
(2) *Surface finish.* The surface finish of part produced with LMD process tends to be rougher than the ones produced by powder bed processes.
(3) *Dimensional accuracy.* Due to the large laser spot size, the melt pool is larger, and thus the dimensional accuracy is not as good as powder bed techniques.

5.4.7 *Applications*

The BeAM's LMD technology can be used in the following areas:

(1) Direct manufacturing
(2) Repairing previously irreparable metal parts
(3) Adding of functions onto existing parts
(4) Producing functionally gradient structures

Fig. 5.12 highlights the repair and adding of functions onto already existed parts using BeAM's LMD technology. The BeAM's LMD system can also be integrated with conventional processes to create unique hybrid manufacturing solutions. For example, BeAM's LMD technology can be used for feature enhancement to an existing component by adding layers of wear-resistant material or other surface treatments.

Figure 5.12: Repair (left) and adding of function (right) onto already existed part with BeAM's LMD technology.
Courtesy of BeAM French Company

5.4.8 *Research and development*

In cooperation with BeAM's ecosystem of partners, BeAM is working consistently on the development of new industrial applications in the field of aeronautic, aerospace, defence and energy. Machines that meet these new demands and requirements are designed and manufactured through an ecosystem of partner community, which includes, but not limited to SAFRAN.

5.5 Arcam's Electron Beam Melting

5.5.1 *Company*

Arcam AB, a Swedish technology development company was founded in 1997. The company's main activity is concentrated in the development of the electron beam melting (EBM) technique for the production of solid metal parts directly from metal powder based on a 3D CAD model. The fundamental development work for Arcam's technology began in 1995 in collaboration with Chalmers University of Technology in Gothenburg, Sweden. The Arcam EBM technology was commercialised in 2001. The address of Arcam AB is Krokslätts Fabriker 27A, SE-431 37 Mölndal, Sweden.

5.5.2 *Products*

Arcam has introduced three main series of products, Arcam Q10 (see Fig. 5.13), Arcam Q20 and Arcam A2X. The Arcam Q10 system is designed to produce orthopaedic implants with its high-productivity and high-resolution features. The Arcam Q20 is targeted at the aerospace market, racing industry and general industry with a machine fulfilling these industries' need for the production of larger components. The Arcam A2X system on the other hand processes materials that require high process temperatures.

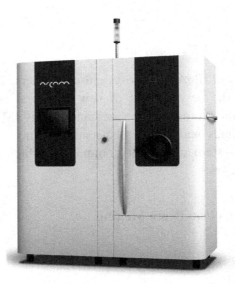

Figure 5.13: Arcam Q10.
Courtesy of Arcam AB

Table 5.6 shows the specifications for the Arcam Q10, Arcam Q20 and Arcam A2X. Currently, four metal powders are available, namely Titanium alloy Ti6Al4V (Grade 5), Titanium alloy Ti6Al4V ELI (Grade 23), Titanium CP (Grade 2) and CoCr Alloy (ASTM F75).

Table 5.6: Specifications of Arcam Q10, Arcam Q20 and Arcam A2X.

Models	Arcam Q10	Arcam Q20	Arcam A2X
Build tank volume	250 × 250 × 200 mm (W × D × H)	420 × 420 mm (Θ × H)	250 × 250 × 400 mm (W × D × H)
Actual build envelope	200 × 200 × 180 mm (W × D × H)	350 × 380 mm (Θ × H)	200 × 200 × 380 mm (W × D × H)
Power supply	3 × 400 V, 32 A, 7 kW		
Size (W × D × H)	Approx. 1850 × 900 × 2200 mm	Approx. 2300 × 1300 × 2600 mm	Approx. 2000 × 1060 × 2370 mm
Weight	1420 kg	2900 kg	1570 kg
Process computer	PC	PC, Windows XP	
CAD interface	Standard: STL		
Network	Ethernet 10/100/1000		
Certification	CE		

5.5.3 *Process*

The Arcam EBM process consists of the following steps:

(1) The part to be produced is first designed in a 3D CAD programme. The model is then sliced into thin layers, approximately a 10th of a millimetre thick.

(2) An equally thin layer of powder is scraped onto a vertically adjustable surface. The first layer's geometry is then created through the layer of powder melting together at those points directed from CAD file with a computer-controlled electron beam.

(3) Thereafter, the building surface is lowered and the next layer of powder is placed on top of the previous layer. The procedure is then repeated so that the object from the CAD model is shaped layer-by-layer until a finished metal part is completed.

5.5.4 *Principle*

The EBM process is based on the following two principles[10]:

(1) Parts are built-up when an electron beam is fired at metal powder. The computer-controlled electron beam in vacuum melts the layer

of powder precisely as indicated by CAD model with the gain in electron kinetic energy.

(2) The building of the part is accomplished layer-by-layer. A layer is added once the previous layer has melted. In this way, the solid details are built-up as thin metal slices melted together.

The basis for the Arcam Technology is essentially EBM. During the EBM process, the electron beam melts metal powder in a layer-by-layer process to build the physical part in a vacuum chamber. The Arcam EBM machines use a powder bed configuration and are capable of producing multiple parts in the same build. The vacuum environment in the EBM machine maintains the chemical composition of the material and provides an excellent environment for building parts with reactive materials like titanium alloys. The electron beam's high power ensures a high rate of deposition and an even temperature distribution within the part, which gives a fully melted metal with excellent mechanical and physical properties.

5.5.5 *Strengths and weaknesses*

The key strengths of the Arcam EBM systems are as follows:

(1) *Superior material properties.* The EBM process produces fully dense metal parts that are void free and have excellent strength and material properties.

(2) *Excellent accuracy.* The vacuum provides a good thermal environment that results in good form stability and controlled thermal balance in the part, greatly reducing shrinkage and thermal stresses. The vacuum environment also eliminates impurities such as oxides and nitrides.

(3) *Excellent finishing.* The high-energy density melting process results in parts that have excellent surface finishing.

(4) *Good build speed.* High-energy density melting with a deflecting electron beam also results in a high-speed build and good power efficiency.

The limitations of the Arcam EBM system are as follows:

(1) *Need to maintain the vacuum chamber.* The process requires a vacuum chamber that has to be maintained as it has direct impact on the quality of the part built.
(2) *High power consumption.* The power consumed for using the electron beam is relatively high.
(3) *Gamma rays.* The electron beam used in the process can produce gamma rays. The vacuum chamber acts as a shield to the gamma rays, thus it is imperative that the vacuum chamber has to be properly maintained.

5.5.6 Applications

(1) *Rapid manufacturing.* Rapid manufacturing includes the fast fabrication of the tools required for mass production, such as specially-shaped moulds, dies and jigs.[11]
(2) *Medical implants.* The technology offers production of small lots as well as customisation of designs, adding important capabilities to the implant industry. A unique feature is the possibility of building parts with designed porosity and scaffolds, which enables the building of implants with a solid core and a porous surface to facilitate bone ingrowth. The ability to directly build complex geometries makes this technology viable for the manufacture of fully functional implants. The process uses standard biocompatible materials such as Ti6Al4V ELI, Ti Grade 2 and cobalt-chrome.
(3) *Aerospace.* This AM system offers significant potential cost savings for the aerospace industry and enables designers to create completely new and innovative systems, applications and vehicles.

5.5.7 Research and development

In 2013, Arcam AB received a European Union research grant to develop a "fast EBM" technology for 3DP. The company is committed to enhance and improve the productivity of the existing EBM technology process by a factor of five with the introduction of new models and new hardware.[12]

5.6 DMG MORI's Hybrid AM

5.6.1 *Company*

DMG MORI AG entered the laser technology sector with the takeover of LCTec GmbH (presently SAUER) in 1999. Since 2001, SAUER represents the advanced technologies within DMG MORI. There are two technology segments in SAUER: ULTRASONIC in Stipshausen (grinding, machining and milling) and LASERTEC in Pfronten (shape, precision tool, fine cutting, power drill and AM). Currently, more than 600 ULTRASONIC and 500 LASERTEC machines had been installed on a worldwide basis. DMG MORI AG is located at Gildemeisterstraße 60, 33689 Bielefeld, Germany.

5.6.2 *Product*

The term hybrid AM can be interpreted as the combination of both additive and subtractive manufacturing in one machine. DMG MORI offers its hybrid AM technology, the LASERTEC 65 *3D* (see Fig. 5.14), by integrating the LDW technology (also known as the LMD process) into its high-tech 5-axis milling machine (previously DMU 65 monoBLOCK®). By combining the flexibility of the LDW process with the precision of the cutting process, it is no longer impossible for the AM technology to produce complete 3D-parts in milling quality. With hybrid AM, the LASERTEC 65 *3D* offers its users with new possibilities of applications and part geometries. Large workpieces with high stock removal volumes can now be machined in an economical way. Equipped with a 2-kW diode laser system, the LASERTEC 65 *3D*'s LDW technology utilises the metal deposition by powder nozzle, enabling building rate of up to 1 kg/h depending on the material used, which is much faster than the powder bed technology. The LASERTEC 65 *3D*

Figure 5.14: LASERTEC 65 *3D*.
Courtesy of DMG MORI

also features flexible changeover between laser and milling operation which allows direct machining of sections which are no longer reachable at the finished part. Table 5.7 shows the specifications for the LASERTEC 65 *3D*.

5.6.3 *Process*

The LASERTEC 65 *3D* is a hybrid machine that combines both AM and milling processes. The AM technology utilised in LASERTEC 65 *3D* is known as the LDW process, which is similar to BeAM's LMD process (refer Section 5.4.3, step (1) and step (2)). The main stages of LASERTEC 65 *3D*'s manufacturing process are as follows:

(1) Slicing of single workpiece sections, as well as the separation of additive and subtractive areas, are done based on the CAD/CAM data.
(2) NC paths are then generated for the laser and milling process in the post-processor.

Table 5.7: Specifications for the LASERTEC 65 *3D*.

Work area	
Traverse path *X/Y/Z*	735/650/560 mm
Work table	
Clamping surface rigid table	800 × 650 mm
NC-swivel-/rotary table	Standard
Dimensions	Ø 650 mm
Max. workpiece weight	600 kg
Rotary axis (C)	360°
Swivel range	±120°
Pmax according to VDI/DGQ 3441	7/9 ws (C-axis/A-axis)
Milling spindle	
RPM (Standard/Option)	10,000/18,000 rpm
Power 40% ED/100% ED (Standard)	13/9 kW
Torque (max.)	83/57 Nm
Tool holder	HSK-A 63
Laser source	
Fibre-coupled diode laser (Standard)	2000 W
Fibre diameter	600 (NA 0.22) μm
Wave length	1030 (± 10) nm
Focal length	200 mm
Diameter laserspot 1 (Standard)	3 mm
Diameter laserspot 2 (Option)	1.6 mm
Building rate (depending on material)	1 kg/h
Powder feeder	
1 supply	Standard
2 supply	Option
Powder tank capacity	5 L
Linear axes (*X/Y/Z*)	
Feed rate	40/40/40 m/min
Rapid traverse rate	40/40/40 m/min
Max. acceleration *X/Y/Z*	6/6/6 m/s^2
Max. feed output (*X/Y/Z*)	7/13/10 kN
Pmax. (*X/Y/Z*) – VDI DGQ 3441	0.008 mm
Tool changer	
Tools Standard/Option	30/60/90
Machine data	
Width × Depth basic machine	4180 × 3487 mm
Machine height	2884 mm
Machine weight	11,300 kg
Control	
CELOS® from DMG MORI with 21.5" ERGOline® and SIEMENS	
DMG ERGOline® Control with Siemens 840D solutionline Operate	

(3) 3D simulation is done with consideration of the integrated laser head to prevent collision.

(4) LDW process and milling process are done on the LASERTEC 65 *3D*. The two processes can be repeated until the intended workpiece is done and the flexible changeover of the two processes is possible (see Fig. 5.15).

(5) Quality inspection is performed on the completed workpiece.

Figure 5.15: Flexible changeover between LDW process (left) and milling process (right).
Courtesy of DMG MORI

5.6.4 *Materials*

LASERTEC 65 *3D* highlights the possibility of multi-material with its LDW technology. The following materials are fully developed to be used in LASERTEC 65 *3D*'s LDW process:

(1) SS 316 L
(2) Inconel 625

Meanwhile, the following materials are still under development to be used in LASERTEC 65 *3D*'s LDW process:

(1) SS (316 L + Si, 304)
(2) Weldable Tool Steels (1.2344 with pre-heating base)

(3)   Bronze and Brass (Cu8Al)
(4)   Cobalt Chrome Molybdenum Alloy
(5)   Inconel 718
(6)   Tungsten Carbide Nickel Alloy

5.6.5 *Principle*

The LASERTEC 65 *3D*'s LDW process is based on the following principles:

(1)   The LDW process welds metal powder to the base material in layers by utilising the laser beam and a coaxial nozzle (non-porous and crack-free melting). With this process, the metal powder is joining up with the surface in a high-strength connection. During the deposition process, a shielding-gas around the nozzle protects against oxidation.
(2)   The metal layers can then be machined mechanically after being cooled down, to achieve desired surface characteristics and dimensions.
(3)   The LASERTEC 65 *3D* system alternates between the milling and laser build process to complete the desired part.

A schematic illustration of the LDW process is shown in Fig. 5.16.

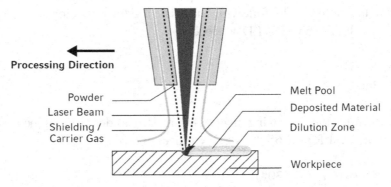

Figure 5.16: Schematic illustration of the LDW process.
Courtesy of DMG MORI

5.6.6 *Strengths and weaknesses*

The key strengths of LASERTEC 65 *3D* system are as follows:

(1) *Hybrid CAD/CAM Module.* Generation of NC-Files for both building and milling of a part can be done in one CAD/CAM system.

(2) *Relatively fast process.* The LASERTEC 65 *3D*'s LDW process is 10 times faster than the powder bed fusion technology.

(3) *Build 3D-contours without support structure.* Machining of a flange or cone can be done without backing geometry.

(4) *The flexible changeover between additive and subtractive processes.* Switching between laser and milling allows for direct finishing of component segments that can no longer be reached on the finished part.

(5) *Possibility of thin-walled structure.* Wall thickness from 0.1 to 5 mm is possible with the LASERTEC 65 *3D* (depending on laser and nozzle geometry).

(6) *Closed-loop control.* Monitoring of the application process and online laser output regulation is done by the integrated optical process monitor.

(7) *Multi-materials are possible.* Two preferred powder feeders can be chosen for the combination of various materials in one part.

The primary limitations of the LASERTEC 65 *3D* system are as follows:

(1) *Lack of fully developed materials.* There are currently only two fully developed materials to be used on LASERTEC 65 *3D*'s LDW process, stainless steel and Inconel 625.

(2) *Limitation of material choices.* The LASERTEC 65 *3D* cannot process aluminium, magnesium, titanium or non-weldable steel.

5.6.7 *Applications*

The LASERTEC 65 *3D* system can be used in the following areas:

(1) *Production.* Produce parts with intricate structures that can hardly be done by using conventional manufacturing processes. Fig. 5.17a shows production of cooling tube.
(2) *Repair.* Previously irreparable parts can be repaired with the LMD technology. Fig. 5.17b shows an impeller which can be repaired using the technology.
(3) *Coating.* Partial or complete coating can be done to shield a part from wear and tear as well as corrosion. Fig. 5.17c shows a coated bearing block using the LASERTEC 65 *3D* hybrid technology.

(a)   (b)   (c)

Figure 5.17: (a) Production of cooling tube, (b) repair of impeller and (c) coating of bearing block.
Courtesy of DMG MORI

5.6.8 *Research and development*

Early models built were often poor in quality and lacked desired physical and mechanical properties, in which they were used mainly for visualisation purposes. Today, many researchers are experimenting with a wider range of materials that are suitable for the intended applications and to improve the physical and mechanical properties of the fabricated models.

5.7 ExOne's Digital Part Materialisation

5.7.1 *Company*

ExOne has its origins with the Extrude Hone Corporation. The late Lawrence J. Rhoades first founded Extrude Hone Corporation and when it expanded and grew, sold the technologies relating to abrasive flow machining, Surftran electrochemical deburring and ThermoBurr deburring to Kennametal, Inc. in 2005. He then set-up ExOne as a new business to serve as an incubator for inventive, new technologies that have the potential to improve manufacturing techniques. Currently, ExOne offers 3D printers that can produce sand and metal pieces in large scales. ExOne's headquarters is located at 127 Industry Boulevard, N. Huntingdon, PA 15642, USA.

5.7.2 *Products*

ExOne focuses on 3DP sand and metal parts with the use of its S-Max+, S-Max, S-Print, M-Print and M-Flex machines. S-Max+ and S-Max machines emphasise on the flexibility and efficiency of the production process while S-Print machines aim to enhance productivity. M-Print machines are designed to achieve the industrial scale of production. Moreover, the M-Flex machine is specialised in producing middle size models and prototypes. ExOne Company's new product — Exerial was also unveiled in 2015, which specialised in series production of intricate sand cores and moulds. Meanwhile, ExOne has the Innovent system, which are customised products designed for researchers and educators. The specifications of the Exerial, S-Max+, S-Max and S-Print machines are summarised in Table 5.8a. The specifications of the M-Print, M-Flex and Innovent systems are summarised in Table 5.8b.

Table 5.8(a): Specifications of the Exerial, S-Max+, S-Max and S-Print machines.

Models	Exerial	S-Max+	S-Max	S-Print
Build volume (L × W × H)	2200 × 1200 × 700 mm × 2 (86.6 × 47.2 × 27.6 in. × 2)	1800 × 1000 × 600 mm (70.9 × 39.4 × 23.6 in.)	1800 × 1000 × 700 mm (70.9 × 39.37 × 27.56 in.)	800 × 500 × 400 mm (29.5 × 15 × 15.75 in.)
Build speed	300,000–400,000 cm³/h (10.6–14.1 ft³/h) (dependent on sand grade)	60,000–85,000 cm³/h (2.12–3.00 ft³/h)	60,000–85,000 cm³/h (2.12–3.00 ft³/h)	Furan: 20,000–36,000 cm³/h (0.71–1.27 ft³/h) Phenolic: 16,000–36,000 cm³/h (0.57–1.27 ft³/h) Silicate: 16,000–18,000 cm³/h (0.57–0.64 ft³/h)
Layer thickness	0.28–0.50 mm (0.011–0.020 in.)	0.28–0.50 mm (0.011–0.020 in.)	0.28–0.50 mm (0.011–0.020 in.)	Furan: 0.28–0.50 mm (0.011–0.02 in.) Phenolic: 0.24 mm (0.009 in.) Silicate: 0.28–0.34 mm (0.011–0.013 in.)
Print resolution	X/Y/Z 0.1/0.1 mm (0.004/0.004 in.)	X/Y/Z 0.1/0.1 mm (0.004/0.004 in.)	X/Y/Z 0.1/0.1 mm (0.004/0.004 in.)	X/Y/Z 0.1/0.1 mm (0.004/0.004 in.)
External dimension (L × W × H)	8380 × 4030 × 4950 mm (329.9 × 158.7 × 194.9 in.)	6900 × 3520 × 2860 mm (271.7 × 138.6 × 112.6 in.)	6900 × 3520 × 2860 mm (271.7 × 138.6 × 112.6 in.)	3270 × 2540 × 2860 mm (128.7 × 100.0 × 112.6 in.)
Weight	12,000 kg (26,456 lbs)	5800 kg (12,787 lbs)	6500 kg (14,330 lbs)	3500 kg (7717 lbs)
Electrical requirements of printer	400 V 3-phase/ N/PE/50-60 Hz, max. 21.5 kW	400 V 3-phase/ N/PE/50–60 Hz, max. 6.2 kW	400 V 3-phase/ N/PE/50–60 Hz, max. 6.3 kW	400 V 3-phase/ N/PE/50–60 Hz, max. 6.2 kW
Electrical requirements of heater	400 V 3-phase/ N/PE/50-60 Hz, max. 12.6 kW	400 V 3-phase/ PE/50-60 Hz, max. 19.2 kW	400 V 3-phase/ PE/50-60 Hz, max. 10.5 kW	400 V 3-Phase/ PE/50–60 Hz, max. 6.3 kW

[a] BAC stands for binder/activator/cleaner.

Table 5.8(a): (*Continued*) Specifications of the Exerial, S-Max+, S-Max and S-Print machines.

Models	Exerial	S-Max+	S-Max	S-Print
Data interface	XPrep	XPrep	STL	STL
Consumable materials	Furan BAC,[a] Silica Sand (280 μm), Ceramic Beads (380 μm)	Phenolic BAC,[a] Ceramic beads, Chromite, Zircon	Furan BAC,[a] Silica sand (280, 380, 500 μm), Black iron oxide, Magnesium inhibitor	Furan: Furan BAC,[a] Silica sand (280, 380, 500 μm), Black iron oxide, Magnesium inhibitor Phenolic: Phenolic BAC,[a] Solvent BAC,[a] Ceramic Beads (280, 380 μm), Chromite, Zircon Silicate: Silicate BAC,[a] Silica sand (280, 380 μm), Ceramic Beads (380 μm)

Table 5.8(b): Specifications of the M-Print, M-Flex and Innovent systems.

Models	M-Print	M-Flex	Innovent
Build volume (L × W × H)	800 × 500 × 400 mm (31.5 × 19.7 × 15.8 in.)	400 × 250 × 250 mm (15.7 × 9.8 × 9.8 in.)	160 × 65 × 65 mm (6.3 × 2.5 × 2.5 in.)
Build speed	Approx. 60 seconds/layer	30–60 seconds/layer (material dependent)	30–60 seconds/layer (material dependent)
Layer thickness	Variable with minimum 0.15 mm (0.006 in.)	Variable with minimum of 0.05 mm (0.002 in.)	Variable with minimum of 0.05 mm (0.002 in.)
Print resolution	X/Y 0.0635 mm (0.003 in.) Z 0.15 mm (0.006 in.)	X/Y 0.0635/ 0.0600 mm (0.003/0.002 in.) Z 0.100 mm (0.004 in.)	X/Y 0.0635/ 0.0600 mm (0.003/0.002 in.) Z 0.100 mm (0.004 in.)
External dimensions (L × W × H)	1675 × 1400 × 1855 mm (66 × 55 × 73 in.)	1675 × 1400 × 1855 mm (66 × 55 × 73 in.)	1203 × 887 × 1434 mm (47.4 × 34.9 × 56.5 in.)

Table 5.8(b): (*Continued*) Specifications of the M-Print, M-Flex and Innovent systems.

Models	M-Print	M-Flex	Innovent
Weight	1020 kg (2249 lbs)	1020 kg (2250 lbs)	320 kg (700 lbs)
Electrical requirements	208–240 V/3 phases	208–240 V/3 phases	120 V 1P 60 Hz/230 V 1P 50 Hz
Data interface	STL, CLI, SLC	STL, CLI, SLC	STL, CLI, SLC
Consumable materials	Binder (aqueous, solvent, phenolic), 420 SS, bronze, thermal support powder	Binder (aqueous, solvent), stainless steel powder (S3, S4), bronze metal powder, thermal support powder, cleaner	Binder (aqueous, solvent), stainless steel powder (S3, S4), bronze powder infiltrant, thermal support powder, cleaner

5.7.3 *Process*

Utilising Digital Part Materialisation (DPM) technology, formerly known as 3DP technology, the process has the ability to build metal components by selectively binding metal powder layer-by-layer with the chemical binder from the digital CAD file. The finished structural skeleton is then sintered and infiltrated to produce a finished part that is 60% dense. The process consists of the following steps. Fig. 5.18 illustrates the DPM process schematically.[13]

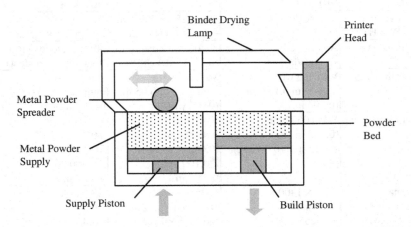

Figure 5.18: Schematic illustration of the DPM process.

(1) A part is first designed on a computer using commercial CAD software.

(2) The CAD image is then transferred to the ExOne System's built-in software. Afterwards, the CAD file is automatically sliced into thin layers (0.1–0.15 mm).

(3) The CAD image is printed with an inkjet print head depositing millions of droplets of binder per second. These droplets dry quickly on deposition.

(4) The process is repeated until the part is completely printed.

(5) The resulting "green" part, of about 60% density, is removed from the machine and excess powder is brushed away.

(6) The "green" part is next sintered in a furnace, while burning off the binder. It is then infiltrated with molten bronze via capillary action to obtain full density. This is carried out in an infiltration furnace.

(7) Post-processing include machining, polishing and coating to enhance wear and chemical resistance, for example nickel and chrome plating.

5.7.4 *Principle*

The working principle of ExOne's DPM uses an electrostatic inkjet printing head to deposit a liquid binder onto the powder metals. The part is built one layer at a time based on sliced cross-sectional data. The metal powder layer is spread on the build piston and a sliced layer is printed onto the powder layer by the inkjet print head depositing droplets of binder that are in turn dried by the binder drying lamp.[14] The process is repeated until the part build is completed.

5.7.5 *Strengths and weaknesses*

The strengths of the DPM process are as follows:

(1) *Fast.* The DPM machine can create multiple parts simultaneously and not sequentially like laser systems. Interchangeable build chambers allow quick turnaround between jobs. Build rates can be up to 85,000 cm^3/h.

(2) *Flexible.* Virtually no restriction of design flexibility, for example complex internal geometries and undercuts can be created.
(3) *Reliable.* There is auto tuning and calibration for maximum performance and also built-in self-diagnostics and status reporting. The inkjet printing process is simple and reliable.
(4) *Large parts.* Large steel mould parts measuring 1800 × 1000 × 700 mm can be produced.

The weaknesses of the DPM process are as follows:

(1) *Large space required.* The machine needs a very large area to house it.
(2) *Limited materials.* The system only prototypes parts with its own metal powder.

5.7.6 *Applications*

The ExOne's DPM is primarily used to rapidly fabricate complex stainless or tool steel tooling parts. Applications include injection moulds, extrusion dies, direct metal components and blow moulding.[15] The technology is also suitable for repairing worn out metal tools.

5.7.7 *Research and development*

ExOne has opened its sixth production service centre and at the time of writing had announced its plan for two new Sales Centre in Brazil and China.[16]

ExOne is investing in non-traditional manufacturing and this promotes the R&D in AM as well as in advancing the micromachining process.

5.8 HP's Multi Jet Fusion™

5.8.1 *Company*

More commonly known as HP, the Hewlett-Packard Company was founded by William "Bill" Redington Hewlett and David "Dave"

Packard in a car garage, situated in Palo Alto. Hewlett-Packard was an American multinational information technology company, which was the world's leading manufacturer of personal computer from 2007 to 2013. At 1st November 2015, HP Inc. became one of the two successors of Hewlett-Packard Company, along with the Hewlett Packard Enterprise. The HP Inc. takes charge of computer and printer products while Hewlett Packard Enterprise focuses on servers, networking, consulting and software products. In October 2014, the Hewlett-Packard Company (currently HP Inc.) announced a new AM technology known as the Multi Jet Fusion (MJF), claiming to be able to enable mass production through AM. HP Inc.'s headquarters is located in Palo Alto, CA, USA.

5.8.2 Products

Following the announcement of the new MJF AM technology in October 2014, HP unveiled two of its MJF machines on 17th May 2016: the HP Jet Fusion 3D 3200 and HP Jet Fusion 3D 4200 (Fig. 5.19). HP highlighted these machines as being able to produce parts at half the cost and up to 10 times faster than fused deposition modelling (FDM) and SLS machines of similar price range. Despite the speed and economic advantage, HP's MJF machines can produce parts with great accuracy and surface finishes. The HP 3200 series was introduced as the entry

Figure 5.19: HP Jet Fusion 3D 4200/3200 Printer.
Courtesy of HP

Table 5.9: Specifications of the HP 3200 series and HP 4200 series.

Models	3200 Series	4200 Series
Printer performance		
Technology	HP MJF technology	
Effective building volume	406 × 305 × 406 mm (16 × 12 × 16 in.)	
Building speed	3500 cm³/h (215 in³/h)	4500 cm³/h (275 in³/h)
Layer thickness	0.08–0.10 mm (0.003–0.004 in.)	0.07–0.12 mm (0.0025–0.005 in.)
Print resolution (*x*, *y*)	1200 dpi	
Dimensions (w × d × h)		
Printer	2178 × 1238 × 1448 mm (85.7 × 48.7 × 57 in.)	
Shipping	2300 × 1325 × 1983 mm (91 × 52 × 78 in.)	
Operating area	3700 × 3700 mm (146 × 146 in.)	
Weight		
Printer	730 kg (1609 lb)	
Shipping	900 kg (1984 lb)	
Network	Gigabit Ethernet (10/100/1000Base-T), supporting the following standards: TCP/IP, DHCP (IPv4 only), TLS/SSL	
Hard disk	2 TB (AES-128 encrypted, FIPS 140, disk wipe DoD 5220 M)	
Software		
Included software	HP SmartStream 3D Build Manager, HP SmartStream 3D Command Center	
Supported file formats	3 MF, STL	
Certified third-party software	Autodesk® Netfabb® Engine for HP, Materialise Magics with Materialise Build Processor for HP MJF	
Power		
Consumption	9–11 kW (typical)	
Requirements	Input voltage three phase 380–415 V (line-to-line), 30 A max, 50/60 Hz/200–240 V (line-to-line), 48 A max, 50/60 Hz	
Certification		
Safety	IEC 60950-1+A1+A2 compliant; United States and Canada (UL listed); EU (LVD and MD compliant, EN60950-1, EN12100-1, EN60204-1 and EN1010)	
Electromagnetic	Compliant with Class A requirements, including: USA (FCC rules), Canada (ICES), EU (EMC Directive), Australia (ACMA), New Zealand (RSM)	
Environmental	RoHS, REACH	
Warranty	One-year Services and Support coverage	

model and is suitable for rapid prototyping. Meanwhile, the HP 4200 series highlights its higher productivity and lower cost per part as compared to the 3200 series, thus making it ideal for short-run production as well as rapid prototyping. Both models come with the HP Jet Fusion 3D Processing Station with Fast Cooling (Fast Cooling is optional for 3200 series), facilitating material recycling and post-processing of the printed part. However, the processing station is sold as a separate unit from the main printer. The specifications of the 3200 Series and 4200 Series are summarised in Table 5.9.

5.8.3 *Process*

The MJF manufacturing process consists of the following steps:

(1) *3D model preparation.* 3D model is prepared and checked with HP's software for errors. Multiple models are then packed into a single build platform and sent to HP's MJF system.

(2) *Material preparation.* Prior to the building process, material cartridges are slotted into HP's processing station where materials are automatically mixed and loaded into HP's build unit. Parts are ready to be built after the build unit is removed from the processing station and slotted into the MJF system. A build unit is a removable unit which contains raw materials before a building process and part cake at the end of a building process.

(3) *Parts building.* After the job data and materials are received by the MJF system, part building can be initiated. Parts are built on a layer-by-layer basis using the MJF technology. The detailed MJF process shall be discussed further in Section 5.8.5.

(4) *Post-processing.* Upon completion of the building process, the build unit containing the part cake can be removed from the MJF system to be processed in the processing unit. A fresh build unit can be slotted into the MJF system to execute its next building job. Meanwhile, the completed part cake has to be cleaned in the processing unit and cooled down before further post-processing.

5.8.4 *Principle*

The MJF technology builds parts on a layer-by-layer basis with the aid of two carriages system as illustrated in Fig. 5.20. The carriage that moves in the *y*-axis is responsible for coating and fusing of materials in the build area, while the carriage that moves in the *x*-axis is responsible for selective "printing" of multiple agents onto the material layer within the build area.

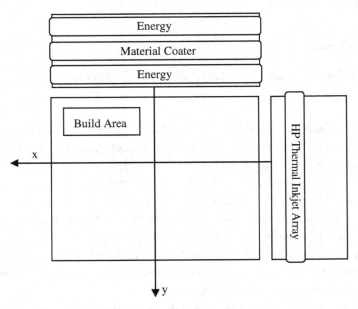

Figure 5.20: Schematic illustration of MJF's carriages system.

A layer of desired part is built by first passing the material coater from top to bottom in the *y*-axis (Fig. 5.20), to deposit a thin layer of material onto the build area (Fig. 5.21a). Then, the HP Thermal Inkjet array prints fusing agent and detailing agent across the build area in one continuous pass. The fusing agent is selectively applied to areas where materials are desired to be fused together after energy exposure (Fig. 5.21b). On the other hand, the detailing agent serves to reduce or amplify the fusing action (Fig. 5.21c). The detailing agent can be used to reduce fusing action on the part boundary where smooth and crisp edges are desired.

The HP Thermal Inkjet Array is capable of printing 30 million drops of agents per second, thereby enabling accurate and precise part building. The layer is then exposed to energy when the carriage (containing material coater and energy depositor) passes from bottom to top in the *y*-axis (Fig. 5.21d). A new layer of materials is recoated at the same time for optimum productivity. The layer that is deposited with fusing agent and later undergoes energy exposure are fused (Fig. 5.21e). The material coating, agent depositing and energy exposure steps are repeated layer after layer until the desired part is completed.

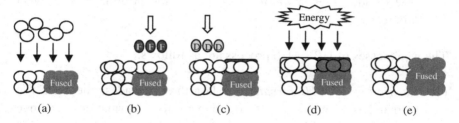

Figure 5.21: (a)–(e) Schematic illustration of MJF's process steps.

5.8.5 *Strengths and weaknesses*

The strengths of HP's MJF process are as follows:

(1) *Fast.* HP claims its systems to be 10 times faster than FDM and SLS machines of similar price range. With many years of experience in 2D thermal inkjet technology, HP is able to deposit fine drops of material precisely at rapid speeds. Instead of selectively depositing materials or energy on one point at a time, HP's MJF systems do so by passing its two carriages system through the build area, thereby reducing the part building time significantly.

(2) *Low cost per part.* HP claims to be able to print parts with a cost that is half of those of other FDM and SLS machines of similar price range. Moreover, HP's current systems feature the PA12 multipurpose thermoplastic which has high reusability of 80%, thus resulting in lesser powder wastage.

(3) *High-quality functional parts.* HP highlights its MJF systems' closed loop thermal control system, monitoring hundreds of points on the powder bed, hence enabling precise thermal control in the build area. The result of this control is high-dimensional accuracy, good precision and optimal mechanical properties of the printed part.

(4) *High accuracy and fine details.* HP's MJF systems can deposit 30 million extremely fine drops of HP fusing and detailing agents per second, allowing highly accurate and precise parts to be produced.

(5) *Add additional parts in the midst of printing.* HP's MJF systems are able to add urgent parts to be printed while printing is already in progress.

The weaknesses of HP's MJF process are as follows:

(1) *Restriction by STL format.* For last three decades, the STL file format has been the de facto standard for information exchange between most AM systems and design software. However, HP's MJF systems could not achieve their maximum potential with STL file format, as this format could not carry the voxel-based information which is required by HP's systems.

(2) *Needs of post-processing.* HP's MJF printed part has to be cleaned from the part cake as well as cooled down before any further processes can proceed.

(3) *Limited variations of materials.* At the point of HP's announcement on its two pioneer MJF models, only the PA12 multipurpose thermoplastic is available for use. Nevertheless, HP planned to offer a wider range of thermoplastics as well as fire retardants and elastomers in the future.

5.8.6 *Research and development*

HP has a plan for the development of MJF technology and is investing in R&D programme on a long-term basis to achieve scalability, modular AM solutions, optimised workflow and advanced materials.

HP's long-term vision on the MJF technology is to enable different and controllable mechanical and physical properties for every single voxel within a single build. HP plans to achieve this vision by utilising the interaction between fusing and detailing agents, and also with the addition of transforming agent to control properties such as translucency and opacity, surface properties, strength and elasticity, colour, electrical conductivity and thermal conductivity.

5.9 Other Notable Powdered-Based AM Systems

There are other AM systems which build parts by utilising technology similar to those discussed in this chapter (Sections 5.1–5.8). These notable powdered-based AM systems shall be discussed as follows:

EOS was founded in 1989 and is the world's leading manufacturer of laser sintering systems today. Based on the laser sintering technology, EOS offers industrial solutions through their AM machines, enabling the possibility to additively manufacture a different spectrum of building materials. The systems FORMIGA P 110, EOS P 396, EOSINT P 760 and EOSINT P 800 are specifically designed for the fabrication of plastic parts. Meanwhile, EOS offers its range of direct metal laser sintering systems, the EOS M 100, EOS M 290, EOS M 400, EOSINT M 280 and PRECIOUS M 080 system to enable AM of metal parts. EOS's headquarters is located at EOS GmbH Electro Optical Systems, Robert-Stiring-Ring 1, D-82152 Krailling, Germany.

Renishaw plc is a British engineering and scientific technology company founded by Sir David McMurtry and John Deer in 1973. It has since become one of the world's leaders in the field of AM, having more than 4000 employees worldwide, with more than 70 offices in 35 different countries. Renishaw's AM systems utilise the powder bed fusion technology to build intricate parts directly from CAD files. Renishaw's range of AM systems includes the AM250, AM400 and RenAM 500M, capable of building metal parts with laser melting process. The company is based in New Mills, Wotton-under-Edge, Gloucestershire, GL12 8JR, United Kingdom.

Concept Laser GmbH (Hofmann Innovation Group) was founded in Lichtenfels, Germany in 2000 with a vision to optimise the SLM process. Concept Laser's AM systems build parts based on the LaserCUSING® technology, where single-component metallic powder materials are fused using a laser. This "generative" method makes it possible to assemble components layer-by-layer from virtually all weldable materials (e.g., stainless steel and hot work steel). Concept Laser's LaserCUSING® product range includes the Mlab cusing/Mlab cusing R, M1 cusing, M2 cusing/M2 cusing Multilaser and X line 2000R. Its address is An der Zeil 8, 96215 Lichtenfels, Germany.

Since 1997, Optomec has focused on commercialising LENS® AM technology for 3D printed metals, which was originally developed by Sandia National Laboratories. Optomec's LENS® systems use proprietary blown powder technology (also known as LMD) to fabricate and repair high-value metal components such as aircraft engine and industrial machinery parts. Optomec offers three models of standard LENS® systems for 3D printed metals: LENS® 450, LENS® MR-7 and LENS® 850-R. Optomec's headquarters is located at 3911 Singer Boulevard, NE, Albuquerque, NM 87109, USA.

Voxeljet AG was incorporated in 1999 by Ingo Ederer under the name Genesis GmbH with the aim of providing new generative processes for production. The special know-how of the company lies in the connection between high-performance inkjet technology and rapid manufacturing. Currently, Voxeljet AG produces six types of AM machines, the VX200, VX500, VX800, VX1000, VX2000 and VX4000. These machines utilise powder binding technology to produce bespoke thermoplastic models from 3D data automatically, without the need for tools. Voxeljet AG's headquarters is located at Paul-Lenz-Straße 1, 86316 Friedberg, Germany.

Since August 1935, Matsuura Machinery Corporation was founded in Fukui city (Japan) as an international heavy machinery manufacturing company. Matsuura's AM technology combines metal laser melting with high-speed milling for the production of highly precise parts. This hybrid AM technology is featured by Matsuura as the LUMEX series. Currently, Matsuura offers two machine types in the LUMEX series, namely Avance-25 and Avance-60. Matsuura's headquarters is located at 1-1 Urushihara-cho Fukui City 910-8530, Japan.

Similar to the SLM technology, 3D systems offers its direct metal printing (DMP) technology to additively manufacture metal parts. 3D systems features six machines in the ProX® DMP series: ProX® DMP 100, ProX® DMP 100 Dental, ProX® DMP 200, ProX® DMP 200 Dental, ProX® DMP 300 and ProX® DMP 320. With a high-precision laser that is capable of melting powdered metal materials completely, the ProX® DMP series builds metal parts in a layer-by-layer basis. 3D system's headquarters is located at 333 Three D Systems Circle, Rock Hill, SC 29730, USA.

References

1. 3D Systems (2007). Products: SLS® Systems. http://www.3dsystems.com/products/sls/index.asp

2. 3D Systems (2015). Selective Laser Sintering (SLS®) Material Selection Guide.

3. Johnson, JL (1994). *Principles of Computer Automated Fabrication*, Chap. 3, pp. 75–84. Irvine: Palatino Press.

4. Sun, MSM, JC Nelson, JJ Beaman and JJ Barlow (1991). A model for partial viscous sintering. In *Proc. of the Solid Freeform Fabrication Symposium*, University of Texas.

5. Hug, WF and PF Jacobs (1991). Laser technology assessment for stereolithographic systems. In *Proc. of the Second International Conference on Rapid Prototyping*, June 23–26, 29–38.

6. Barlow, JJ, MSM Sun and JJ Beaman (1991). Analysis of selective laser sintering. *Proc. of the Second International Conference on Rapid Prototyping*, June 23–26, 29–38.

7. 3D Systems (DTM Corp) (1999). *Horizons Q4*, pp. 6–7.

8. Prototype Magazine (2007). *MCP SLM Machines*. http://www.prototypemagazine.com/index.php?option=com_content&task= view&id=102&Itemid=2.

9. Van Elsen, M, F Al-Bender and J Krith (2008). Application of dimensional analysis to selective laser melting. *Rapid Prototyping Journal*, 14(1), 15–22.

10. Arcam AB (2007). Electron Beam Melting. http://www.arcam.com/technology/tech_ebm.asp.

11. Gibbons, GJ and RG Hansell (2005). Direct tool steel injection mold inserts through the Arcam EBM free-form fabrication process. *Assembly Automation*, 25(4), 300–305.

12. Arcam AB Nears Completion of Fast EBM 3D Printing Research. Engineering.com. http://www.engineering.com/3DPrinting/3DPrintingArticles/ArticleID/6062/Arcam-AB-Nears-Completion-of-Fast-EBM-3D-Printing-Research.aspx.

13. The Editors (2001). Added options for producing "impossible" shapes, rapid traverse—Technology and trends spotted by the Editors of Modern Machine Shop. *MMS Online*.

14. Vasilash, GS (2001). A quick look at rapid prototyping. *Automotive Design and Production*.

15. Waterman, PJ (2000). RP3: Rapid prototyping, pattern making and production. *DE Online*.

16. ExOne. *ExOne Opens Sixth Production Service Centre*. http://www.exone.com/en/news/exone-opens-sixth-production-service-center.

Problems

1. Using a sketch to illustrate your answer, describe the SLS process.

2. Discuss the types of materials available for the SLS Series.

3. Describe the main differences between the CJP process and the SLS process.

4. List the advantages and disadvantages of the CJP process.

5. Discuss possible applications for SLM process.

6. Both SLM and BeAM's LMD processes utilise laser to build metal parts. How does LMD process distinguish itself from the SLM process?

7. Describe Arcam's EBM technology.

8. List the advantages and disadvantages of the EBM process.

9. What are the critical factors that influence the performance and applications of the following AM processes?

    (a) 3D Systems' SLS,

    (b) 3D Systems CJP,

    (c) BeAM's LMD,

    (d) Arcam's EBM

10. Name three laser powder-based AM systems and three non-laser powder-based AM systems.

11. What does the term "Hybrid" mean in DMG MORI's Hybrid AM system? How does the hybrid process overcome some disadvantages of the LMD process?

12. Discuss the advantages and disadvantages of powder-based AM systems compared with:

    (a) liquid-based AM systems,

    (b) solid-based AM systems

13. List the significances of ExOne's DPM process.

14. Discuss the differences between ExOne's DPM process and that of the SLS process for the production of metal parts.

15. List ExOne's field of operation for their AM system.

16. Describe the process and principles of HP's MJF process. What is the purpose of detailing agent in the MJF process?

CHAPTER 6

ADDITIVE MANUFACTURING
DATA FORMATS

6.1 STL Format

Representation methods used to describe computer-aided design (CAD) geometry vary from one system to another. A standard interface is needed to convey geometric descriptions from various CAD packages to additive manufacturing (AM) systems. For the last three decades, the **ST**ereo**L**ithography (STL) file format, as the *de facto* standard, has been used in many, if not all, AM systems to exchange information between design programs and AM systems.[1]

The STL file,[1-3] conceived by 3D Systems, USA, is created from the CAD database via an interface on the CAD system. This file consists of an unordered list of triangular facets representing the outside skin of an object. There are two STL file formats. One is the American Standard Code for Information Interchange (ASCII) format and the other is the binary format. The size of ASCII STL file is larger than that of the binary format, but is human-readable. In an STL file, triangular facets are described by a set of X, Y and Z coordinates for each of the three vertices and a unit normal vector with X, Y and Z to indicate the side of the facet, which is inside or outside the object. An example is shown in Fig. 6.1.

As the STL file is a facet model derived from precise CAD drawings, it is an approximate model of the part. Moreover, many commercial CAD models are not robust enough to generate the facet model (STL file) and frequently have problems as a result.

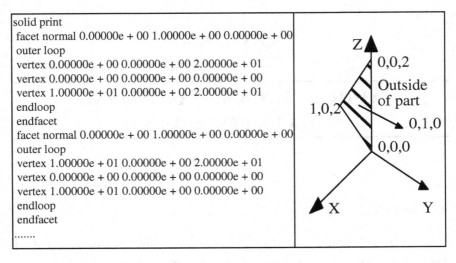

```
solid print
facet normal 0.00000e + 00 1.00000e + 00 0.00000e + 00
outer loop
vertex 0.00000e + 00 0.00000e + 00 2.00000e + 01
vertex 0.00000e + 00 0.00000e + 00 0.00000e + 00
vertex 1.00000e + 01 0.00000e + 00 2.00000e + 01
endloop
endfacet
facet normal 0.00000e + 00 1.00000e + 00 0.00000e + 00
outer loop
vertex 1.00000e + 01 0.00000e + 00 2.00000e + 01
vertex 0.00000e + 00 0.00000e + 00 0.00000e + 00
vertex 1.00000e + 01 0.00000e + 00 0.00000e + 00
endloop
endfacet
.......
```

Figure 6.1: A sample STL file.

Nevertheless, there are several advantages of the STL file. First, it provides a simple method of representing three-dimensional (3D) CAD data. Second, it is already a *de facto* standard and has been used by most CAD and AM systems. Finally, it can provide small and accurate files for data transfer.

On the other hand, several disadvantages of the STL file exist. First, the STL file is many times larger than the original CAD data file for a given accuracy parameter. The STL file carries much redundant information such as duplicate vertices and edges, as shown in Fig. 6.2. Second, geometry flaws exist in STL files because many commercial tessellation algorithms used by CAD vendors today are not sufficiently robust. This gives rise to the need for a "repair software", which slows the production cycle time. Third, the STL file carries limited information to represent colour, texture material, substructure and other properties of the manufactured end object. Finally, the subsequent slicing of large STL files can take many hours. However, some AM processes can slice while they are building the previous layer, and this will alleviate this disadvantage.

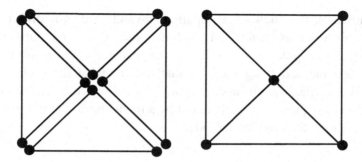

Figure 6.2: Edge and vertex redundancy in STL format.

6.2 STL File Problems

Several problems plague STL files and they are due to the very nature of STL files as they contain no topological data. Many commercial tessellation algorithms used by CAD vendors today are also not robust,[4–6] and as a result they tend to create polygonal approximation models which exhibit the following types of errors:

(1) Gaps (cracks, holes and punctures), that is, missing facets
(2) Degenerate facets (where all its edges are collinear)
(3) Overlapping facets
(4) Non-manifold topology conditions

The underlying problem is due, in part, to the difficulties encountered in tessellating trimmed surfaces, surface intersections and controlling numerical errors. This inability of the commercial tessellation algorithm to generate valid facet model tessellations makes it necessary to perform model validity checks before the tessellated model is sent to the AM equipment for manufacturing. If the tessellated model is invalid, procedures become necessary to determine what the specific problems are, whether they are due to gaps, degenerate facets or overlapping facets and so forth.

Early research has shown that repairing invalid models is difficult and not at all obvious.[7] However, before proceeding any further into discussing the procedures that can be generated to resolve these difficulties, the following sections shall clarify what the problems, as mentioned earlier, are. In addition, an illustration will be presented to show the consequences brought about by a model having a missing facet, that is, a gap in the tessellated model.

6.2.1 *Missing facets or gaps*

Tessellation of surfaces with large curvature can result in errors at the intersections between such surfaces, leaving gaps or holes along the edges of the part model.[8] A surface intersection anomaly which results in a gap is shown in Fig. 6.3.

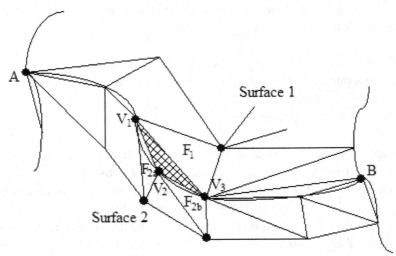

Figure 6.3: Gaps due to missing facets.[4]

6.2.2 Degenerate facets

A geometrical degeneracy of a facet occurs when all of the facets' edges are collinear even though all its vertices are distinct. This might be caused by stitching algorithms that attempt to avoid shell punctures as shown in Fig. 6.4(a).[9]

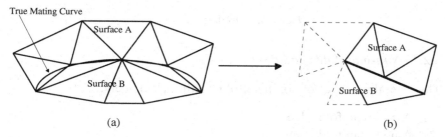

Figure 6.4: Shell punctures (a) created by unequal tessellation of two adjacent surface patches along their common mating curve and (b) eliminated at the expense of adding a degenerate facet.

The resulting facets generated, shown in Fig. 6.4(b), eliminate the shell punctures. However, this is done at the expense of adding a degenerate facet. While degenerate facets do not contain valid surface normals, they do represent implicit topological information on how two surfaces are mated. This important information is consequently stored prior to discarding the degenerate facet.

6.2.3 Overlapping facets

Overlapping facets may be generated due to numerical round-off errors that occur during tessellation. The vertices are represented in 3D space as floating point numbers instead of integers. Thus, the numerical round-off can cause facets to overlap if tolerances are set too liberally. An example of an overlapping facet is illustrated in Fig. 6.5.

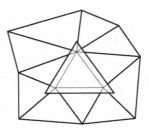

Figure 6.5: Overlapping facets.

6.2.4 *Non-manifold conditions*

There are three types of non-manifold conditions, namely:

(1) A non-manifold edge
(2) A non-manifold point
(3) A non-manifold face

These may be generated because tessellations of fine features are susceptible to round-off errors. An illustration of a non-manifold edge is shown in Fig. 6.6(a). Here, the non-manifold edge is actually shared by four different facets, as shown in Fig. 6.6(b). A valid model would be one whose facets have only an adjacent facet each, that is, one edge shares two facets only. Hence, the non-manifold edges must be resolved such that each facet has only one neighbouring facet along each edge, that is, by reconstructing a topologically manifold surface.[4] Shown in Fig. 6.6(c) and (d) are two other types of non-manifold conditions.

All problems that have been mentioned previously are difficult for most slicing algorithms to handle, and cause fabrication problems for AM processes which essentially require valid tessellated solids as input. Moreover, these problems arise because tessellation is a first-order approximation of more complex geometric entities. Thus, such problems have become almost inevitable as long as solid models are represented using the STL format, which inherently has these limitations.

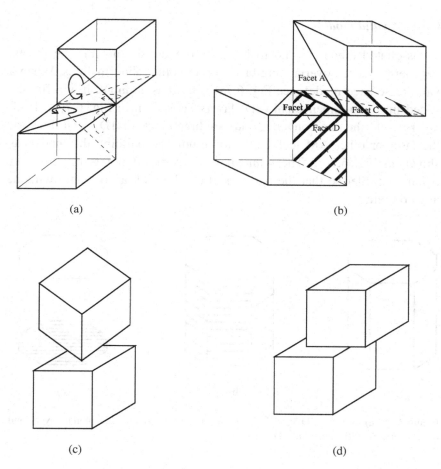

(a)   (b)

(c)   (d)

Figure 6.6: A non-manifold edge (a) whereby two imaginary minute cubes share a common edge and (b) whereby four facets share a common edge after tessellation. (c) Non-manifold point and (d) non-manifold face.

6.3 Consequences of Building a Valid and Invalid Tessellated Model

Each of the following sections presents an example of the outcome of a model built using a valid and invalid tessellated model, when used as an input to AM systems.

6.3.1 *A valid model*

A tessellated model is said to be valid if there are no missing facets, degenerate facets, overlapping facets or any other abnormalities. When a valid tessellated model (see Fig. 6.7a) is used as an input, it will first be sliced into two-dimensional (2D) layers, as shown in Fig. 6.7(b). Each layer would then be converted into unidirectional (or 1D) scan lines for the laser or other AM techniques to commence building the model, as shown in Fig. 6.7(c). The scan lines act as on/off points for the laser beam controller so that the part model can be built accordingly without any problems.

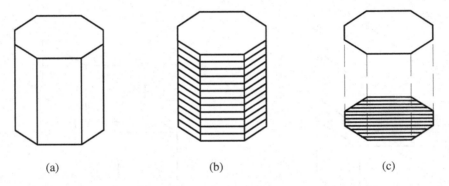

(a)   (b)   (c)

Figure 6.7: (a) A valid 3D model, (b) a 3D model sliced into 2D planar layers and (c) conversion of 2D layers into 1D scan lines.

6.3.2 *An invalid model*

However, if the tessellated model is invalid, a situation may develop as shown in Fig. 6.8.

A solid model is tessellated non-robustly and results in a gap as shown in Fig. 6.8(a). If this error is not corrected and the model is subsequently sliced, as shown in Fig. 6.8(b), in preparation for it to be built layer by layer, the missing facet in the geometrical model would cause the system to have no predefined stopping boundary on the particular slice. Thus, the building process would continue right to the physical limit of the AM machine, creating a stray physical solid line and ruining the part being produced, as illustrated in Fig. 6.8(c). Therefore, it is of utmost importance that the model is "repaired" before it is sent for building. Thus, the model validation and repair problem are stated as follows:

Given a facet model (a set of triangles defined by their vertices), in which there are gaps, i.e., missing one or more sets of polygons, generate 'suitable' triangular surfaces which 'fill' the gaps.[4]

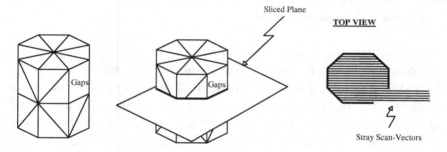

Figure 6.8: (a) An invalid tessellated model, (b) an invalid model being sliced and (c) a layer of an invalid model being scanned.

6.4 STL File Repair

The STL file repair can be implemented using a generic solution and dedicated solutions for special cases.

6.4.1 *Generic solution*

In order to ensure that the model is valid and can be robustly tessellated, one solution is to check the validity of all the tessellated triangles in the model. This section presents the basic problem of missing facets and a proposed generic solution to solve the problem.

In existing AM systems, when a punctured shell is encountered, the course of action taken usually requires a skilled technician to manually repair the shell. This manual shell repair is frequently done without any knowledge of the designer's intent. The work can be very time-consuming and tedious, thus negating the advantages of AM as the cost would increase and the time taken might be longer than that taken if traditional prototyping processes were used.

The main problem of repairing the invalid tessellated model would be that of matching the solution to the designer's intent when it may have been lost in the overall process. Without knowledge of the designer's intent, it would indeed be difficult to determine what the "right" solution should be. Hence, an "educated" guess is usually made when faced with ambiguities of the invalid model.

The algorithm for a generic solution to solve the "missing facets" problem aims to match, if not exceed, the quality of repair done manually by a skilled technician when information of the designer's intent is not available. The basic approach of the algorithm would be to detect and identify the boundaries of all the gaps in the model. Once the boundaries of the gap are identified, suitable facets would then be generated to repair and "patch up" these gaps. The size of the generated facets would be restricted by the gap's boundaries while the orientation of its normal would be controlled by comparing it with the rest of the shell. This is to ensure that the generated facet orientation is correct and consistent throughout the gap closure process.

The orientation of the shell's facets can be obtained from the STL file, which lists its vertices in an ordered manner following Mobius' rule. The algorithm exploits this feature so that the repair carried out on the invalid model, using suitably created facets, would have the correct orientation. Thus, this generic algorithm can be said to have the ability to make an inference from the information contained in the STL file so that the following two conditions can be ensured:

(1) The orientation of the generated facet is correct and compatible with the rest of the model.
(2) Any contoured surface of the model would be followed closely by the generated facets due to the smaller facet generated. This is in

contrast to manual repair whereby, in order to save time, fewer facets generated to close the gaps are desired, resulting in large generated facets that do not follow closely to the contoured surfaces.

Finally, the basis for the working of the algorithm is due to the fact that in a valid tessellated model, there must only be two facets sharing every edge. If this condition is not fulfilled, then this indicates that there are some missing facets. With the detection and subsequent repair of these missing facets, the problems associated with an invalid model can then be eliminated.

6.4.1.1 *Solving the "missing facets" problem*

The following procedure illustrates the detection of gaps in the tessellated model and its subsequent repair. It is carried out in four steps:

Step 1: Checking for approved edges with adjacent facets

The checking routine executes as follows for Facet A, as seen in Fig. 6.9:

(a)   i. Read in first edge {vertex 1-2} from the STL file.
    ii. Search file for a similar edge in the opposite direction {vertex 2-1}.
    iii. If edge exists, store this under a temporary file (e.g., file B) for approved edges.

(b)   i. Read in second edge {vertex 2-3} from the STL file.
    ii. Search file for a similar edge in the opposite direction {vertex 3-2}.
    iii. Perform as in *a(iii)* above.

(c)   i. Read in third {vertex 3-1} from the STL file.
    ii. Search file for a similar edge in the opposite direction {vertex 1-3}.
    iii. Perform as in *a(iii)* above.

This process is repeated for the next facet until all the facets have been searched.

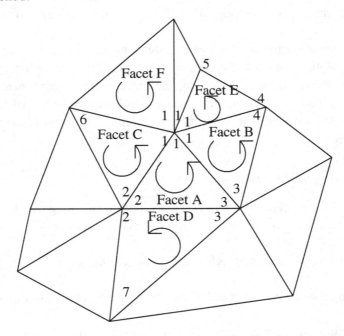

Figure 6.9: A representation of a portion of a tessellated surface without any gaps.

Step 2: Detection of gaps in the tessellated model

With reference to Fig. 6.10, the detection routine executes as follows:

(a)   i.   For Facet A, read in edge {vertex 2-3} from the STL file.
       ii. Search file for a similar edge in the opposite direction {vertex 3-2}.
       iii. If edge does not exist, store edge {vertex 3-2} in another temporary file (e.g., file C) for suspected gap's bounding edges and store vertex 2-3 in file B1 for existing edges without adjacent facets (this would be used later for checking the generated facet orientation).

(b) i. For Facet B, read in edge {vertex 5-2} from the STL file.
   ii. Search file for a similar edge in the opposite direction {vertex 2-5}.
   iii. If it does not exist, perform as in a(iii) above.

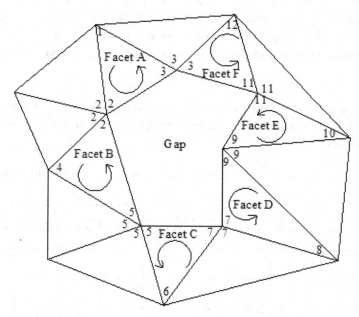

Figure 6.10: A representation of a portion of a tessellated surface with a gap.present.

(c) i. Repeat for edges: 5-2; 7-5; 9-7; 11-9; 3-11.
   ii. Search for edges: 2-5; 5-7; 7-9; 9-11; 11-3.
   iii. Store all the edges in the temporary file B1 for edges without any adjacent facet and store all the suspected bounding edges of the gap in temporary file C. File B1 can appear as in Table 6.1.

Table 6.1: File B1 contains existing edges without adjacent facets.

Edge						
Vertex	First	Second	Third	Fourth	Fifth	Sixth
First	2	7	3	5	9	11
Second	3	5	11	2	7	9

Step 3: Sorting of erroneous edges into a closed loop

When the checking and storing of edges (both with and without adjacent facets) are completed, a sort would be carried out to group all the edges without adjacent facets to form a closed loop. This closed loop would represent the gap detected and be stored in another temporary file (e.g., file D) for further processing. The following is a simple illustration of what could be stored in file C for edges that do not have an adjacent edge:

Assuming all the "erroneous" edges are stored according to the detection routine (see Fig. 6.10 for all the erroneous edges), then file C can appear as in Table 6.2.

Table 6.2: File C containing all the "erroneous" edges that would form the boundary of each gap.

Edge								
Vertex	First	Second	Third	Fourth	Fifth	Sixth	Seventh	Eighth	Ninth
First	3	5	*	11	2	7	*	9	*
Second	2	7	*	3	5	9	*	11	*

* Represents all the other edges that would form the boundaries of other gaps.

As can be seen in Table 6.2, all the edges are unordered. Hence, a sort would have to be carried out to group all the edges into a closed loop. When the edges have been sorted, it would then be stored in a temporary file, say file D. Table 6.3 is an illustration of what could be stored in file D.

Table 6.3: File D containing sorted edges.

Edge					
Vertex	First	Second	Third	Fourth	Fifth	Sixth
First	3	2	5	7	9	11
Second	2	5	7	9	11	3

Fig. 6.11 is a representation of the gap, with all the edges forming a sorted closed loop.

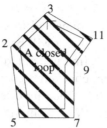

Figure 6.11: A representation of a gap bounded by all the sorted edges.

Step 4: Generation of facets for the repair of the gaps

When the closed loop of the gap is established with its vertices known, facets are generated one at a time to fill up the gap. This process is summarised in Table 6.4 and illustrated in Fig. 6.12.

Table 6.4: Process of facet generation.

		V3	V2	V5	V7	V9	V11
Generation of facets	F1	1	2	–	–	–	3
	F2	E	1	–	–	2	3
	F3	E	1	2	–	3	E
	F4	E	E	1	2	3	E

V = vertex, F = facet, E = eliminated from the process of facet generation.

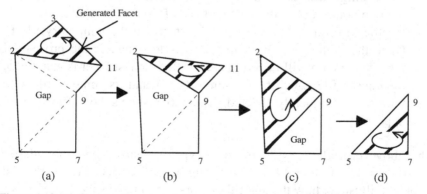

Figure 6.12: (a) First facet generated, (b) second facet generated, (c) third facet generated and (d) fourth facet generated.

With reference to file D,

(1) Generating the first facet: First two vertices (V3 and V2) in the first two edges of file D will be connected to the first vertex in the last edge (V11) in file D, and the facet will be stored in a temporary file E (see Table 6.5 on how the first generated facet would be stored in file E). The facet is then checked for its orientation using the information stored in file B1. Once its orientation is determined to be correct, the first vertex (V3) from file D will be temporarily removed.

(2) Generating the second facet: Of the remaining vertices in file D, the previous second vertex (V2) will become the first edge of file D. The second facet is formed by connecting the first vertex (V2) of the first edge with that of the last two vertices in file D (V9, V11), and the facet is stored in temporary file E. It is then checked to confirm if its orientation is correct. Once it is determined to be correct, the vertex (V11) of the last edge in file D is then removed temporarily.

(3) Generating the third facet: The whole process is repeated as it was done in the generation of facets 1 and 2. The first vertex of the first two edges (V2, V5) is connected to the first vertex of the last edge (V9) and the facet is stored in temporary file E. Once its orientation is confirmed, the first vertex of the first edge (V2) will be removed from file D temporarily.

(4) Generating the fourth facet: The first vertex in the first edge will then be connected to the first vertices of the last two edges to form the fourth facet and it will again be stored in the temporary file E. Once the number of edges in file D is less than 3, the process of facet generation will be terminated. After the last facet is generated, the data in file E will be written to file A and its content (file E's) will be subsequently deleted. Table 6.5 shows how file E may appear.

The above procedures work for both types of gaps whose boundaries consist either of an odd or even number of edges. Fig. 6.13 and Table 6.6 illustrate how the algorithm works for an *even* number of edges or vertices in file D.

Table 6.5: Illustration of how data could be stored in file E.

Generated	First Edge		Second Edge		Third Edge	
Facet	First vertex	Second vertex	First vertex	Second vertex	First vertex	Second vertex
First	V3	V2	V2	V5	V5	V3
Second	V2	V9	V9	V11	V11	V2
Third	V2	V5	V5	V9	V9	V2
Fourth	V5	V7	V7	V9	V9	V5

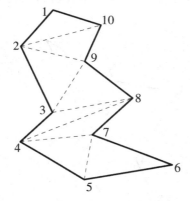

Figure 6.13: Gaps with an even number of edges.

Table 6.6: Process of facet generation for gaps with even number of edges.

Facets	Vertices									
	V1	V2	V3	V4	V5	V6	V7	V8	V9	V10
F1	1	2								3
F2	E	1							2	3
F3	E	1	2						3	E
F4	E	E	1					2	3	E
F5	E	E	1	2				3	E	E
F6	E	E	E	1			2	3	E	E
F7	E	E	E	1	2		3	E	E	E
F8	E	E	E	E	1	2	3	E	E	E

With reference to Table 6.6,

First facet generated: Second facet generated:
Edge 1 → V1, V2   Edge 1 → V2, V9
Edge 2 → V2, V10   Edge 2 → V9, V10
Edge 3 → V10, V1   Edge 3 → V10, V1

and so on until the whole gap is covered. Similarly, Fig. 6.14 and Table 6.7 illustrate how the algorithm works for an *odd* number of edges or vertices in file D.

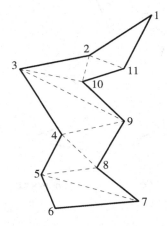

Figure 6.14: Gaps with odd number of edges.

Table 6.7: Process of facet generation for gaps with odd number of edges.

Facets	Vertices										
	V1	V2	V3	V4	V5	V6	V7	V8	V9	V10	V11
F1	1	2									3
F2	E	1								2	3
F3	E	1	2							3	E
F4	E	E	1						2	3	E
F5	E	E	1	2					3	E	E
F6	E	E	E	1				2	3	E	E
F7	E	E	E	1	2			3	E	E	E
F8	E	E	E	E	1		2	3	E	E	E
F9	E	E	E	E	1	2	3	E	E	E	E

The process of facet generation for *odd* vertices is also done in the same way as *even* vertices. The process of facet generation has the following pattern:

F1 → First and second vertices are combined with the last vertex. Once completed, eliminate first vertex. The remainder is 10 vertices.

F2 → First vertex is combined with last two vertices. Once completed, eliminate the last vertex. The remainder is nine vertices.

F3 → First and second vertices are combined with the last vertex. Once completed, eliminate first vertex. The remainder is eight vertices.

F4 → First vertex is combined with last two vertices. Once completed, eliminate the last vertex. The remainder is seven vertices.

This process is continued until all the gaps are patched.

6.4.1.2 *Comparison with an existing algorithm for facet generation*

An illustration of an existing algorithm that might cause a very narrow facet (shaded) to be generated is shown in Fig. 6.15.

This results from using an algorithm that uses the smallest angle to generate a facet. In essence, the problem is caused by the algorithm's search for a time-local rather than a global-optimum solution.[10] Also, calculation of the smallest angle in 3D space is very difficult.

Fig. 6.15 is similar to Fig. 6.14. However, in this case, the facet generated (shaded) can be very narrow. In comparing the algorithms, the result obtained would match, if not, exceed the algorithm that uses the smallest angle to generate a facet.

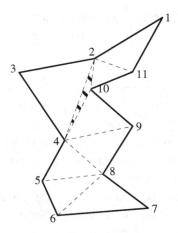

Figure 6.15: Generation of facets using an algorithm that uses the smallest angle between edges.

6.4.2   *Special algorithms*

The generic solution presented could only cater to gaps (whether simple or complex) that are isolated from one another. However, should any of the gaps meet at a common vertex, the algorithm may not be able to work properly. In this section, the algorithm is expanded to include solving some of these special cases. The special cases include:

(1)   Two or more gaps are formed from a coincidental vertex
(2)   Degenerate facets
(3)   Overlapping facets

The special cases are classified as such because these errors are not commonly encountered in the tessellated model. Hence, it is not advisable to include this expanded algorithm in the generic solution as it can be very time-consuming to apply during a normal search. However, if there are still problems in the tessellated model after the generic solution's repair, the expanded algorithm can then be used to detect and solve the special case problems.

6.4.2.1 *Overlapping facets*

The condition of overlapping facets can be caused by errors introduced by inconsistent numerical round-off. This problem can be resolved through vertex merging where vertices within a predetermined numerical round-off tolerance of one another can be merged into just one vertex. Fig. 6.16 illustrates one example of how this solution can be applied. Fig. 6.17 illustrates another example of an overlapping facet.

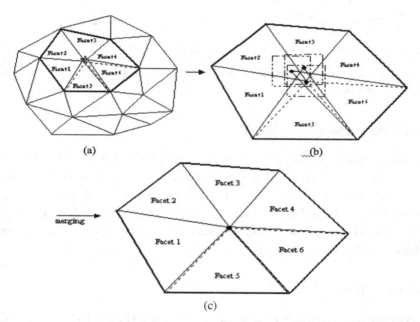

(a)   (b)

(c)

Figure 6.16: (a) Overlapping facets, (b) numerical round-off equivalence region and (c) vertices merged.

It is recommended that this merging of vertices be done before the searching of the model for gaps. This will eliminate unnecessary detection of erroneous edges and save substantial computational time expended in checking whether the edges can be used to generate another facet.

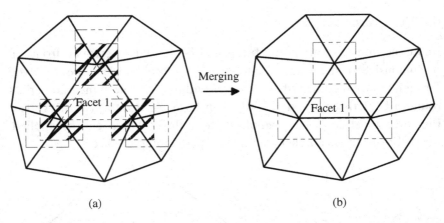

Figure 6.17: (a) An overlapping facet and (b) facet's vertices merged with vertices of neighbouring facets.

6.4.3 *Performance evaluation*

Computational efficiency is an issue whenever CAD model repair of solids that have been finely tessellated is considered. This is due to the fact that for every unit increase in the number of facets (finer tessellation), the additional increase in the number of edges is three. Thus, the computational time required for the checking of erroneous edges would correspondingly increase.

6.4.3.1 *Efficiency of the detection routine*

Assuming that there are 12 triangles in the cube (Fig. 6.18), the number of edges = $12 \times 3 = 36$. The number of searches is computed as follows:

(1) Read first edge, search 35 edges and remove two edges.
(2) Read second edge, search 33 edges and remove two edges.
(3) Read third edge, search 31 edges and remove two edges and so on.

$$\text{Number of searches} = 1 + 3 + \ldots + 35$$

In general,

$$\text{Number of searches} = 1 + ... + (n - 1)$$
$$= n^2/4 \qquad (6.1)$$

where n = number of edges

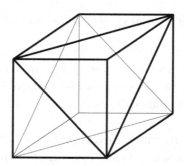

Figure 6.18: A cube tessellated into 12 triangles.

Although this result does not seem satisfactory, it is both an optimum and a robust solution that can be obtained given the inherent nature of the STL format (such as its lack of topological information).

However, if topological information is available, the efficiency of the routine used to detect the erroneous edges can definitely be increased significantly. Some additional points worth noting are: first, binary files are far more efficient than ASCII files because they are only 20%–25% as large and thus reduce the amount of physical data that need to be transferred because they do not require the subsequent translation into a binary representation. Consequently, one easily saves several minutes per file by using binary instead of ASCII file formats.[10]

As for vertex merging, the use of one-dimensional AVL-tree can significantly reduce the search-time for sufficiently identical vertices.[10] The AVL-trees, which are usually twelve to sixteen levels deep, reduces each search from 0(n) to 0(log n) complexity, and the total search-time ranges from close to an hour to less than a second.

6.4.3.2 *Estimated computational time for shell closure*

The computational time required for shell closure is relatively fast. The estimated time can range from a few seconds to less than a minute and is arrived at based on the processing time obtained by Jan Helge Bohn[10] that uses a similar shell closure algorithm.

6.4.3.3 *Limitations of current shell closure process*

The shell closure process developed thus far does not have the ability to detect or solve the problems posed by any of the non-manifold conditions. However, the detection of non-manifold conditions and their subsequent solutions would be the next focus in ongoing research.

A limitation of the algorithm involves the solving of coplanar (see Fig. 6.19a) and non-coplanar facets (see Fig. 6.19b) whose intersections result in another facet. The reasons for such errors are related to the application that generated the faceted model, the application that generated the original 3D CAD model and the user.

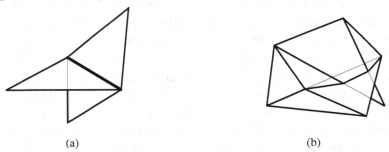

       (a)   (b)

Figure 6.19: (a) Incorrect triangulation (coplanar facet) and (b) non-coplanar whereby facets are split after being intersected.

Another limitation involves the incorrect triangulation of parametric surface (see Fig. 6.20). One of the overlapping triangles, $T_b = BCD$ should not be present and should thus be removed while the other triangle, $T_a = ABC$ should be split into two triangles so as to maintain the correct contoured surface. The proposed algorithm is presently unable to solve this problem.[11]

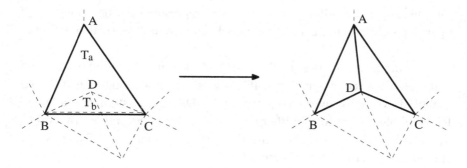

Figure 6.20: Incorrect triangulation of parametric surface.

Finally, as mentioned earlier, the efficiency of $n^2/4$ (when n is the number of edges) is a major limitation especially when the number of facets in the tessellated model becomes very large (e.g., greater than 40,000). There are differences in the optimum use of computational resources. However, further work is being carried out to ease this problem by using topological information that is available in the original CAD model.

6.5 Other Translators

Many translators have been researched upon and proposed.[12-14] Others are proposed based on existing formats and these are given in the next sections.

6.5.1 *IGES file*

Initial Graphics Exchange Specification (IGES) is a standard used to exchange graphics information between commercial CAD systems. It was set up as American National Standard in 1981.[15,16] The IGES file can precisely represent CAD models. It includes not only the geometry information (Parameter Data Section) but also topological information (Directory Entry Section). In the IGES, surface modelling, constructive solid geometry (CSG) and boundary representation (B-rep) are introduced. Especially, the ways of representing the regularised operations for union, intersection, and difference have been also defined.

The advantages of the IGES standard are its wide adoption and comprehensive coverage. Since IGES was set up as American National Standard, virtually every commercial CAD/CAM system has adopted IGES implementations. Furthermore, it provides the entities of points, lines, arcs, splines, NURBS surfaces and solid elements. Therefore, it can precisely represent CAD models.

However, several disadvantages of the IGES standard in relation to its use as an AM format include the following objections:

(1) Because IGES is the standard format to exchange data between CAD systems, it also includes much redundant information that is not needed for AM systems.
(2) The algorithms for slicing an IGES file are more complex than the algorithms slicing a STL file.
(3) The support structures needed in AM systems such as SLA cannot be created according to the IGES format.

IGES is generally used as a data transfer medium, which interfaces with various CAD systems. It can precisely represent CAD models. Advantages of using IGES over current approximate method include precise geometry representations, few data conversions, smaller data files and simpler control strategies. However, the problems are lack of transfer standards for variety of CAD systems and system complexities.

6.5.2   *HP/GL file*

Hewlett-Packard Graphics Language (HP/GL) is a standard data format for graphic plotters.[1,2] Data types are all 2D, including lines, circles, splines, texts and so forth. The approach, as seen from a designer's point of view, would be to automate a slicing routine that generates a section slice, invoke the plotter routine to produce a plotter output file and then loop back to repeat the process.

The advantages of the HP/GL format are that a lot of commercial CAD systems have the interface to output HP/GL format and it is a 2D geometry data format which does not need to be sliced.

However, there are two distinct disadvantages of the HP/GL format. First, because HP/GL is a 2D data format, the files would not be appended, potentially leaving hundreds of small files needing to be given logical names and then transferred. Second, all the support structures required must be generated in the CAD system and sliced in the same way.

6.5.3 *CT data*

Computerised tomography (CT) scan data is a particular approach for medical imaging.[1,17] This is not a standardised data. Formats are proprietary and somewhat unique from one CT scan machine to another. The scan generates data as a grid of 3D points, where each point has a varying shade of grey indicating the density of the body tissue found at that particular point. Data from CT scans have been used to build skull, femur, knee, and other bone models on STL systems. Some of the reproductions were used to generate implants, which have been successfully installed in patients. The CT data consist essentially of raster images of the physical objects being imaged. It is used to produce models of human temporal bones.

There are three approaches to make models out of CT scan information:

(1) Via CAD systems
(2) STL-interfacing
(3) Direct interfacing

The main advantage of using CT data as an interface of AM is that it is possible to produce structures of human body by the AM systems. However, some disadvantages of CT data include: first, the increased difficulty in dealing with image data as compared with STL data and second, the need for a special interpreter to process CT data.

6.6 Standards for Representing Additive Manufactured Objects

6.6.1 *AM file format*

For decades, research and development has been going on for the AM industry to specify an international standard. As AM technology experienced fast and furious changes from producing primarily homogeneous products of simple geometry to producing heterogeneous objects of complex shapes, textures and colours, the industry standard for representing AM objects has evolved.[18] In 2011, the two most prominent standard developers worldwide, American Society for Testing and Materials (ASTM) International and the International Standards Organisation (ISO) signed a Partner Standards Developing Organisation (PSDO) cooperative agreement to establish common standards based on the work of the ASTM International Committee F42 and ISO Technical Committee 261 on AM.[19] The ASTM International Committee F42 on AM Technologies was formed in 2009 to develop standards which would play a critical role in all aspects of the AM industry[20] while the ISO/TC 261 was created in 2011. The F42 committee has published several standards regarding terminology, AM file format (AMF) and part production of several metals using AM technologies. In the current manufacturing environment, the standards adopted by the industry are categorised under two groups: informal standards and formal standards.

STL format is grouped under informal standards commonly called industry or *de facto* standards. For the several past decades in the AM industry, STL format is the most common interface between computer design programs and AM systems due to its simplicity and ease of use (see Section 6.1). However, as STL format only contains information on the surface mesh and has no provisions for representing colour, texture, material, substructure and even more complex 3D shapes using multiple materials are created by an expanding AM industry, new standard interchange file formats have to be developed to deal with these problems and address the needs.

The ISO 10303 series, a group of standards widely known as STandard for the Exchange of Product model data (STEP), is considered under the formal category. They are approved internationally and supported by a public standards-making authority for the electronic exchange of product data between computer-based product life cycle systems under the ISO Technical Committee ISO/TC 184/SC 4 on industrial data.[21] STEP covers a broad range of different product types and life cycle stages. STEP internally consists of several application protocols (APs), which are used to describe a particular life cycle stage of a particular product type. The exchange of the actual information indicating description and the method used is purely based on APs. To implement the description, a set of integrated resources (IRs) are constructed to form the APs and the construction of the IRs is significantly applied for specific purposes. Developers have long noticed the importance of tackling the emerging problems due to the unprecedented scale of the ISO 10303 standard. To track the new evolvements in technology while they are occurring, STEP is currently subjected to revision. One example would be the strong emphasis on the use of extensible markup language (XML) as a means of exchanging STEP information.

AMF format Version 1.1 is the new standard issued under the collaboration between ASTM and ISO in 2013 which answers the growing need of a standard interchange file format that can provide detailed properties of the target product. The new standard, ISO/ASTM 52915:2013, will greatly benefit the whole AM industry, allowing CAD designers, scanners and 3D graphical editors to conveniently work with AM equipment.[18] Similar to the STL format, the AMF format is based on triangles and may be sorted, expressed and transferred via paper or electronic means, as long as the data required by this specification are collected. When prepared in a systematic electronic format, an XML, a standard which provides for tagging of information content within documents, must be closely followed in order to support standards-compatible interoperability.

AMF format aims to address the following aspects[18]:

Technology independence — AMF allows any AM machines to build the part to the best of their capability without using exclusive technologies.

Simplicity — AMF allows easy implementation and understanding. The format can be displayed and rectified in common ASCII text viewers.

Scalability — AMF could tackle increasing in part complexity and size and with the improving resolution and precision of manufacturing machines.

Performance — AMF allows appropriate duration for read-and-write process and moderate file size.

Backwards compatibility — Interchangeable between existing STL file and new AMF file.

Future compatibility — AMF caters to future changes, allowing new features to be amounted to the parts using the old machines.

AMF has its significant advantage in that it is more flexible. As the required information is stored in XML format that is a widely accepted ASCII text file, amendments could be made to the file as long as the add-on adheres to the XML standard.[18,22] Moreover, XML is both machine and human readable which offers convenience in coding and debugging.

The general structure of an AMF file begins with the XML declaration line specifying the XML version and encoding. Blank lines and standard XML comments can be interspersed in the file and these will be ignored by the typical interpreter. The main file is enclosed between an opening </amf> element and a closing </amf> element. These elements are necessary to denote the file type, as well as to fulfil the requirement that all XML files have a single-root element. The unit system can also be specified (mm, inch, ft, meters or micrometres). In the absence of a unit specification, millimetres are assumed.

Within the AMF brackets, there are five top-level elements and these include:

\<object>

The object element specifies the geometry and it defines a volume or volumes of material, each of which is associated with a material identification (ID) for printing.

The top-level \<object> element specifies a unique ID and contains two child elements: \<vertices> and \<volume>. The required \<vertices> element lists all vertices that are used in this object. Each vertex is implicitly assigned a number in the order in which it was declared starting at zero. The required child element \<coordinates> gives the position of the point in 3D space using the \<x>, \<y> and \<z> elements.

After the vertex information, at least one \<volume> element shall be included. Each volume encapsulates a closed volume of the object. Multiple volumes can be specified in a single object. While volumes may share vertices at interfaces they may not overlap. In each volume, the child element \<triangle> shall be used to define triangles that tessellate the surface of the volume. Each \<triangle> element will comprise of three vertices from the set of indices of the previously defined vertices. The indices of the three vertices of the triangles are specified using the \<v1>, \<v2> and \<v3> elements according to the right-hand rule such that vertices are listed in counterclockwise order as viewed from outside the volume. Each triangle is implicitly assigned a number in the order in which it is declared starting from zero.

Note that the file does not describe support structure, only the intended object to be manufactured.

Only a single object element is required for a fully functional AMF file; the other top-level elements are optional.

\<material\>

The optional material element defines one or more materials for printing with an associated material ID. If no material element is included, a single default material is assumed.

Each material is assigned a unique ID. Geometric volumes are associated with materials by specifying a material ID within the \<volume\> element. The material ID "0" is reserved for no material (void).

Material attributes are contained within each \<material\>. The element \<colour\> is used to specify the red/green/blue/alpha (RGBA) values in a specified colour space for the appearance of the material. It can be inserted at the material level to associate a colour with a material, the object level to colour an entire object, the volume level to colour an entire volume, a triangle level to colour a triangle or a vertex level to associate a colour with a particular vertex. Object colour overrides material colour specification, a volume colour overrides an object colour, vertex colours override volume colours and triangle colouring overrides a vertex colour.

Additional material properties can be specified using the \<metadata\> element, such as the material name for operational purposes or elastic properties for equipment that can control such properties.

\<texture\>

The optional texture element defines one or more images or textures for colour or texture mapping each with an associated texture ID.

\<constellation\>

The optional constellation element hierarchically combines objects and other constellations into a relative pattern for printing. If no constellation elements are specified, each object element will be imported with no relative position data.

A constellation can specify the position and orientation of objects to increase packing efficiency and describe large arrays of identical objects. The <instance> element specifies the displacement and rotation an existing object in order for it to be specified in its position in the constellation. The displacement and rotation are always defined relatively to the original position and orientation in which the object was originally defined. Rotation angles are specified in degrees and are applied first to rotation about the x axis, then the y axis and then the z axis.

A constellation can refer to another constellation, with multiple levels of hierarchy. Recursive or cyclic definitions of constellations, though, are not permissible.

<metadata>

The optional metadata element specifies additional information about the object(s) geometries, and materials being defined in the file. Information can include a name, textual description, authorship, copyright information and special instructions.

The AMF file can be stored either in plain text or be compressed. If compressed, the compression shall be in ZIP archive format (3) and can be done manually or at write time using any one of several open compression libraries. Both the compressed and uncompressed version of this file will have the AMF extension, and it is the responsibility of the parsing program to determine whether or not the file is compressed and if so, to perform decompression during read.

6.6.2 *3D manufacturing format*

Apart from AMF, there is another new 3D printing format launched by the 3D manufacturing format (3MF) Consortium in 2015.[23] 3MF allows design applications to send full-fidelity 3D models to a host of other applications, platforms, services and printers. The 3MF format is designed in an AM format, with the complete model information contained within a single archive: mesh, textures, materials, colours and print ticket. 3MF provides a clear definition of manifoldness — open source code available for rapid validation and there is no ambiguity for models with self-intersections. Transforms and object references are supported, with multiple objects contained within the single archive. Single objects can be referenced or moved without changing the mesh and multiple identical objects can be placed referencing the same mesh.

A 3MF document is an open packaging conventions (OPC) package that holds a 3D payload and follows the rules developed by the 3MF Consortium. The 3MF document format includes a well-defined set of parts and relationships, each fulfilling a particular purpose in the document.[23] A typical 3MF document is illustrated in Fig. 6.21.

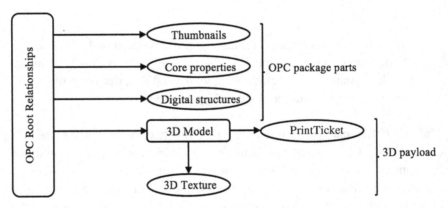

Figure 6.21: A typical 3MF document.

6.6.3 *3D payload*

A payload is a complete collection of interdependent parts and relationships within a package. A payload that has a 3D Model root part is known as a 3D payload. There can be more than one 3D payload in a 3MF document, but there is only one primary 3D payload. A specific relationship type is defined to identify the root of a 3D payload within a 3MF document: the 3MF document StartPart relationship. The primary 3D payload root is the 3D Model part that is referenced by the 3MF document StartPart relationship to find the primary 3D payload in a package. The 3MF document StartPart relationship must point to the 3D Model part that identifies the root of the 3D payload. The payload includes the full set of parts required for processing the 3D Model part. Parts included to the 3D payload are explicitly linked to the 3D payload root by relationship. 3MF documents must not reference resources external to the 3MF document package unless specified otherwise in an extension.[23]

The 3D Model part contains definitions of one or more objects to be fabricated by the 3D manufacturing processes. The 3D Model part is the only valid root of a 3D payload. A 3D Model part has two sections: a set of resource definitions that include objects and materials, as well as a set of specific items to actually build.

Thumbnails are small images that represent the contents of an entire 3MF document. Thumbnails enable external agents to view the contents of the 3MF document easily.

PrintTicket parts provide user intent and device configuration information to printing consumers. Only one PrintTicket part can be attached to a 3D Model part. The PrintTicket format is governed by the specific consumer environment.

6.6.4 *3MF document markup*

3MF document markup has been designed to facilitate independent development of compatible systems that produce or consume 3MF documents. The design of 3MF document markup reflects the trade-offs between two, sometimes competing, goals[23]:

(1) 3MF document markup should be parsimonious; that is, it should include only the minimum set of primitive operations and markup constructs necessary to manufacture common 3D objects with full fidelity. Redundancy in the specification increases the

opportunity for independent implementations to introduce accidental incompatibilities. Redundancy also increases the cost of implementation and testing, and, typically, the required memory component.

(2) 3MF document markup should be compact; that is, the most common primitives should have compact representations. Bloated representations compromise the performance of systems handling 3MF documents. As byte-count increases, so does communication time. Although compression can be used to improve communication time, it cannot eliminate the performance loss caused by bloated representations.

3MF document markup has been designed in anticipation of the evolution of this specification. It also allows third parties to extent the markup. Advanced features are built as extensions, using an A La Carte model whereby producers can state explicitly which extensions are used (by declaring the matching XML namespace in the <model> element) and consumers can state explicitly which extensions they support, so other tools in the chain know which parts will be ignored. Versioning is accomplished concurrently, as the namespace will be updated to reflect a version change. Therefore, versioning happens independently for the core specification and for each extension, and the version of each can be determined by checking its namespace.

Some of the commonly used elements in 3MF include[23]:

<model>

The <model> element is the root element of the 3D Model part. There must be exactly one <model> element in a 3D Model part. A model may have zero or more child metadata elements. A model must have two additional child elements: <resources> and <build>. The <resources> element provides a set of definitions that can be drawn for to define a 3D object. The <build> element provides a set of items that should actually be manufactured as part of the job.

\<metadata\>

Metadata in 3MF document without a namespace must be restricted to names and values defined by this specification. If a name value is not defined in this specification, it must be prefixed with the namespace name of an XML namespace declaration on the \<model\> element that is not drawn from the default namespace.

\<resources\>

The \<resources\> element acts as the root element of a library of constituent pieces of the overall 3D object definition. Objects, properties and materials are collectively referred to as resources in this specification. Each resource might rely on other resources for its complete definition. For example, an object resource may refer to material resources or even other object resources to fully describe a 3D object. An object resource represents a single 3D object that could be manufactured, but not necessarily will be manufactured. The objects that actually will be manufactured are referenced from an \<item\> element child of the \<build\> element. Objects are defined as resources primarily to aid in modularising design and re-use of component, thus compacting the overall markup size.

\<build\>

The \<build\> element contains one or more items to manufacture as part of processing the job. A consumer must not output any 3D objects not referenced by an \<item\> element.

\<item\>

The \<item\> element identifies one object resource to be output by the 3D manufacturing device. A consumer must apply the transform prior to outputting the object. A 3MF document may include multiple objects to manufacture at the same time. The arrangement of these items in the build is considered a default; consumers may rearrange the items for manufacturing in order to better pack the build volume. Sometimes objects are arranged in the coordinate space so as to be manufactured in an interlocking fashion; producers of these objects should collect them as components, as 3D manufacturing devices must not transform components of an object relative to each other. 3D manufacturing devices should avoid overlapping the items when packing the build volume.

<object>

An object resource is defined by an <object> element. An <object> element has attributes for the property group and specific property member that are to be applied to the entire object, except where overridden by a descendant element, such as <triangle> element or a component-referenced <object> element. If this object contains any triangles with assigned materials, the object must specify pid and pindex, to act as default values for any triangles with unspecified properties. If no properties are assigned at all, the choice for the properties of the object is left to the consumer.

Part numbers are intended as a way to keep track of objects which may have been modified during a tool chain. When editing or processing a 3MF document, these part numbers should be preserved to the greatest degree possible, duplicating them for objects split into pieces, removing them for objects that are combined and maintaining them for objects that are modified.

<mesh>

The <mesh> element is the root of a triangular mesh representation of an object volume. It contains a set of vertices and a set of triangles. If the mesh is under an object of the type "model", it must have:

Manifold edges: Every triangle edge in the mesh shares common vertex endpoints with the edge of exactly one other triangle.

Consistent triangle orientation: Every pair of adjacent triangles within the mesh must have the same orientation of the face normal towards the exterior of the mesh, meaning that the order of declaration of the vertices on the shared edge must be in the opposite order.

Outward-facing normals: All triangles must be oriented with normals that point away from the interior of the object. Meshes with negative volume will not be printed (or will become voids). In combination with the preceding two rules, a mesh is therefore a continuous surface without holes, gaps, open edges or non-orientable surfaces (e.g., Klein bottle).

<vertex>

A <vertex> element represents a point in 3D space that is referenced by a triangle in the mesh. The decimal values representing the coordinates can be recorded to arbitrary precision. Producers should not use more precision than the error generated in their calculations, or the anticipated resolution of their consumer. The variable-precision nature of ASCII encoding is a significant advantage over fixed-width binary formats, and helps make up the difference in storage efficiency.

The <vertices> element contains all the <vertex> elements for this object. The vertices represent the corners of each triangle in the mesh. The order of these elements defines an implicit o-based index that is referenced by other elements, such as the <triangle> element. The producer should not include duplicate vertices unless coalescing duplicates would create non-manifold edges. Furthermore, a producer should collapse vertices that are very closely proximal with a single vertex whenever appropriate. In order to avoid integer overflows, a vertex array must contain less than 2^{31} vertices.

\<triangles\>

A \<triangle\> element represents a single face of the mesh. The order of the vertices (v1, v2 and v3) must be specified in counter-clockwise order, such that the face normal of the triangle is pointing towards the outside of the object. The indices v1, v2 and v3 must be distinct. The properties applied to each vertex (p1, p2 and p3) allow property gradients to be defined across the triangle, where interpolation of the property is defined as the linear convex combination. A consumer that cannot create property gradients must apply the p1 property to the entire triangle. A consumer that cannot use properties on a per-triangle basis must ignore the triangle properties and use the \<object\> level property instead. If p1 is not specified then no properties are assigned to the triangle. If p2 or p3 is unspecified then p1 is used for the entire triangle. The property group is specified by the pid attribute, if different than the property group defined at the object level. Since this is applied to the whole triangle, it implicitly forces the three properties to be from the same group, which implies they are of the same type, as defined by possible extensions to this specification.

If the properties defined on the triangle are from a \<basematerials\> group, they must not form gradients, as interpolation of base materials is not defined in this core specification. Therefore, p1, p2 and p3 must be equal or unspecified. Material gradients and interpolation methods are defined in extension specifications.

The \<triangles\> element contains a set of one or more \<triangle\> elements to describe a full 3D object mesh. If the object type is "model", the mesh has to contain at least four triangles to form a solid body. In order to avoid integer overflows, a triangle array MUST contain less than 2^{31} triangles.

\<component>

A component selects a predefined object resource and adds it to the current object definition, after applying the provided matrix transform. This composition of an object definition from multiple primitive components can provide a very compact file size for a quite complex model. In keeping with the use of a simple parser, producers must define objects prior to referencing them as components.

The \<components> element acts as a container for all components to be composed into the current object. A component is an object resource that is used in the context of another object definition. Through the use of components, a producer can reduce the overall size of the 3MF document. A 3D manufacturing device must respect the relative positions of the component objects; it must not transform them relative to each other except as specified in the document. In order to avoid integer overflows, a components element must contain less than 2^{31} components.

\<base>

A base material is used to define the specific material to be used for manufacturing certain objects in a model. In particular, support objects are often built from a different material than the non-sacrificial portion of the model. Since these materials can be applied at both the object and triangle level, they are technically only specifying the material at the surface of the object. Consumers may choose how the materials are distributed through the volume, so long as the surfaces have the specified materials.

Base material names are intended to convey design intent and producers should avoid machine-specific naming in favor of more portable descriptions. Printer-specific information is intended to live in the PrintTicket, which includes a mapping from these base

Table 6.8: Comparison between AMF and 3MF.

Item	AMF	3MF
Specification	ISO/ASTM 52915:2013(E) (Version 1.1)	3MF Core Specification (Version 1.0) 3MF Materials and Property Extension (Version 1.0)
Data representation	XML based	XML based
Data storage	Plain text or compressed ZIP archive	ZIP archive only (i.e., more compact)
Support multi-material, multi-colour, multi-objects, 2D and 3D texture, metadata	Yes	Yes
Support curved triangles and edges	Yes	Not specified
Support scripting for graded materials, porous materials and stochastic materials	Yes	Not specified
Support tolerance, surface roughness, coating, support structures, etc.	Future features	Not specified
Support digital signature and content protection	Future features	Yes
Support subtractive manufacturing devices	Not specified	Yes
Support older STL files	Need conversion tools	Yes
Support Windows print pipeline	Not specified	Yes
Maturity of file format	Mature (work with printers in the market today)	Under development (Soon to become mature)

materials to actual printer materials. Base material names should be unique throughout the 3MF package.

A <basematerials> element is a material group that acts as a container for the base materials. The order of these elements forms an implicit 0-based index that is referenced by other elements, such as the <object> and <triangle> elements. Other types of property group elements can be added as extensions to this specification, due to the <any> element under resources. These groups allow different types of properties to be separated and organised, given the many possible extensions.

3MF file format has the following benefits:

(1) It is rich enough to fully describe a model, retaining internal information, colour and other characteristics.
(2) It is extensible so that it supports new innovations in 3D printing.
(3) It is interoperable and is able to be broadly adopted.
(4) It is free of the issues besetting other widely used file formats.

6.6.5 *Comparison between AMF and 3MF*

3MF is also XML-based and features geometry representation similar to AMF, but in a more compact and size-friendly format. The file defines all standard, optional and mandatory parts, with complete model information contained in a single archive. The comparison between the two formats is summarised in Table 6.8.

6.7 Other Standards on AM

There are several international standards that are being developed and published for AM.

The ISO/ASTM 52921:2013: Standard Terminology for AM — Coordinate Systems and Test Methodologies is a standard for terminology that describes terms, definitions of terms, descriptions of terms, nomenclature and acronyms associated with coordinate systems and testing methodologies for AM technologies. This is for AM users, producers, researchers, educators, press/media and others to have a common understanding of these terminology, particularly when reporting results from the testing of parts made on AM systems.

The ASTM F2792-12a: Standard Terminology for AM Technologies is an ASTM international standard that formalised the terms, definitions of terms, descriptions of terms, nomenclature and acronyms associated with AM technologies to provide standardised terminology for usage by AM users, producers, researchers, educators, press/media and others.

The ASTM F2971 — 13: Standard Practice for Reporting Data for Test Specimens Prepared by AM describes a standard procedure for reporting results by testing or evaluation of specimens produced by AM. This practice provides a common format for presenting data for AM specimens, for: (a) to establish further data reporting requirements and (b) to provide information for the design of material property databases.

The ASTM F2924 — 14: Standard Specification for AM Titanium-6 Aluminium-4 Vanadium with Powder Bed Fusion covers additively manufactured titanium-6aluminum-4vanadium (Ti-6Al-4V) components using full-melt powder bed fusion such as electron beam melting and laser melting. The components produced by these processes are used typically in applications that require mechanical properties similar to machined forgings and wrought products. Components manufactured to this specification are often, but not necessarily, post processed via machining, grinding, electrical discharge machining (EDM), polishing and so forth to achieve desired surface finish and critical dimensions.

The ASTM F3001 — 14: Standard Specification for AM Titanium-6 Aluminium-4 Vanadium ELI (Extra Low Interstitial (ELI)) with Powder Bed Fusion establishes the requirements for additively manufactured Ti-6Al-4V with ELI components using full-melt powder bed fusion such as electron beam melting and laser melting. The standard covers the classification of materials, ordering information, manufacturing plan, feedstock, process, chemical composition, microstructures, mechanical properties, thermal processing, hot isostatic pressing, dimensions and mass, permissible variations, retests, inspection, rejection, certification, product marking and packaging, and quality program requirements.

The ASTM F3056 — 14: Standard Specification for AM Nickel Alloy (UNS N06625) with Powder Bed Fusion covers additively manufactured nickel alloy components using full-melt powder bed fusion such as electron beam melting and laser melting. The components produced by these processes are used typically in applications that require mechanical properties similar to machined forgings and wrought products. Components manufactured to this specification are often, but not necessarily, post processed via machining, grinding, EDM, polishing and so forth to achieve desired surface finish and critical dimensions. This specification is intended for the use of purchasers or producers, or both, of additively manufactured nickel alloy components for defining the requirements and ensuring component properties.

There are other standards that are under discussion at ISO/ASTM technical committees and there will be more of these standards published as AM becomes more popular and before widely used in the industry.

References

1. Jacobs, PF (1992). *Rapid Prototyping and Manufacturing.* Dearborn, MI: Society of Manufacturing Engineers.

2. Famieson, R and H Hacker (1995). Direct slicing of CAD models for rapid prototyping. *Rapid Prototyping Journal,* ISATA94, Aachen, Germany.

3. Donahue, RJ (1991). CAD model and alternative methods of information transfer for Rapid Prototyping Systems. In *Proceedings of the Second International Conference on Rapid Prototyping,* pp. 217–235. Dayton, OH, June 23–26.

4. Wozny, MJ (1992). Systems issues in Solid Freeform Fabrication. In *Proc. Solid Freeform Fabrication Symposium 1992,* pp. 1–5. Texas, USA.

5. Leong, KF, CK Chua and YM Ng (1996). A study of stereolithography file errors and repair Part 1 — Generic solutions. *International Journal of Advanced Manufacturing Technologies,* 12(6), 407–414.

6. Leong, KF, CK Chua and YM Ng (1996). A study of stereolithography file errors and repair Part 2 — Special cases. *International Journal of Advanced Manufacturing Technologies,* 12(6), 415–422.

7. Rock, SJ and MJ Wozny (1991). A flexible format for solid freeform fabrication. In *Proceedings, Solid Freeform Fabrication Symposium 1991*, pp. 1–12. Texas, USA.

8. Crawford, RH (1993). Computer aspects of solid freeform fabrication: Geometry, process control and design. In *Proceedings, Solid Freeform Fabrication Symposium 1993*, pp. 102–111. Texas, USA.

9. Bohn, JH and MJ Wozny (1992). Automatic CAD-model repair: Shell-closure. In *Proceedings, Solid Freeform Fabrication Symposium 1992*, pp. 86–94. Texas, USA.

10. Bohn, JH (1993). *Automatic CAD-Model Repair.* Ann Arbor, MI: UMI.

11. Dolenc, A and I Malela (1992). A data exchange format for LMT processes. In *Proceedings of the Third International Conference on Rapid Prototyping*, pp. 4–12. Dayton, USA.

12. Chua CK, JGK Gan and M Tong (1997). Interface between CAD and rapid prototyping systems part I: A study of existing interfaces. *International Journal of Advanced Manufacturing Technology*, 13(8), 566–570.

13. Chua CK, GKJ Gan and M Tong (1997). Interface between CAD and rapid prototyping systems part II: LMI — An improved interface. *International Journal of Advanced Manufacturing Technology*, 13(8), 571–576.

14. Gan, J, CK Chua and M Tong (1999). Development of new rapid prototyping interface. *Computers in Industry*, 39(1), 61–70.

15. Reed, K, D Harrvd and W Conroy (1990). *Initial Graphics Exchange Specification (IGES) Version 5.0.* CAD-CAM Data Exchange Technical Centre.

16. Jinghon, L (1992). Improving stereolithography parts quality — Practical solutions. In *Proceedings of the Third International Conference on Rapid Prototyping*, pp. 171–179.

17. Swaelens, B and JP Kruth (1993). Medical applications of rapid prototyping techniques. In *Proceedings of the Fourth International Conference on Rapid Prototyping*, pp. 107–120.

18. *ISO ASTM International, Standard Specification for Additive Manufacturing File Format (AMF) Version 1.1*, ISO/ASTM 52915:2013(E), ISO, Case postate 56,

CH-1211, Geneva 20, Switzerland, and ASTM International, 100 Barr Harbor Drive, PO Box C700, West Conshohocken, PA 19428-2959, US.

19. ASTM International, Standardisation News. (2013). *Additive Manufacturing to Benefit from Standards Agreement: ASTM and ISO Sign Additive Manufacturing PSDO Agreement.* http://www.astm.org/standardisation-news/outreach/ astm-and-iso-sign-additive-manufacturing-psdo-agreement-nd11.html

20. ASTM International, Technical Committees. (2013). *Committee F42 on Additive Manufacturing Technologies.* http://www.astm.org/CO MMITTEE/F42.htm

21. ISO Standards Development, Technical Committee. (2013). *ISO/TC 184/SC 4 Industrial data.* http://www.iso.org/iso/home/standards_development/ list_of_iso_ technical_committees/iso_technical_committee.htm?commid=54158

22. The World Wide Web Consortium. (2014). *Extensible Markup Language (XML) 1.0*, 5th Ed. http://www.w3.org/TR/REC-xml/

23. *3D Manufacturing Format Specification and Reference Guide.* http://3mf.io/ wp-content/uploads/2016/03/3MFcoreSpec_1.1.pdf

Problems

1. What is the *de facto* industry format adopted by AM systems for the past three decades? Describe the format and illustrate with an example. What are the pros and cons of using this format?

2. Referring to Fig. 6.22, write a sample STL file for the shaded triangle.

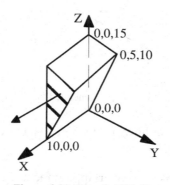

Figure 6.22: Sample STL file.

3. Based on the STL format, how many triangles and coordinates would a cube contain?

4. What causes *missing facets or gaps* to occur?

5. Illustrate, with diagrams, the meaning of *degenerate facets*.

6. Explain *overlapping facets*.

7. What are the three types of non-manifold conditions?

8. What are the consequences of building a valid and invalid tessellated model?

9. What problems can the generic solution solve?

10. Describe the algorithm used to solve the *missing facets* problem.

11. In Fig. 6.23, facet X is incorrectly oriented. Describe how the problem can be resolved. Draw the newly generated facet X with the corrected orientation.

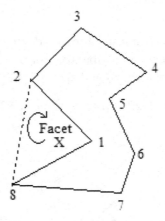

Figure 6.23: Incorrectly generated facet's orientation.

12. Prove and illustrate how a degenerate facet can be repaired by the use of vector algebra.

13. How can the problem of overlapping facets be solved?

14. What is the efficiency of the detection routine? Illustrate using the example of a cube.

15. What are some of the limitations of the solutions, both generic and special cases, used to solve STL-related problems?

16. Name some other translators used in place of STL.

17. What is the new ASTM International standard file format for AM systems? Discuss its advantages and disadvantages compared to STL file format.

18. What are the five top elements within the AMF file?

19. Discuss some of the future features that might be developed based on the AMF at the current state.

20. Discuss the advantages of 3MF over AMF.

CHAPTER 7

APPLICATIONS AND EXAMPLES

7.1 Application — Material Relationship

The applications of additive manufacturing (AM) are closely related to the purposes of prototyping and manufacturing, and consequently the materials used. In other words, if the materials used in AM are similar to the materials traditionally used for prototyping or manufacturing in their mechanical properties, the range of applications will be very wide. Unfortunately, there are marked differences in the mechanical properties of current AM materials and traditional manufacturing materials. The key to increasing the applicability of AM technologies, therefore, lies in widening the range of materials that can be additively manufactured.

In the early development of AM systems, the emphasis of the tasks at hand was oriented towards the creation of "visualisation" and "touch-and-feel" models to support design — that is, to create 3D objects with little or no regard to their function and performance. These are broadly classified as "Applications in Design". This was influenced and in many cases limited by the materials available in these early AM systems. However, vendors were also continually on the look-out for more areas of applications, such as functional evaluation and testing, and ultimately tooling applications. This called not only for improvements in strength and accuracy provided by AM technologies, but also the development of an even wider range of materials, including metals, ceramics and composites. Applications of AM prototypes were first extended to "Applications in Engineering, Analysis and Planning" and later extended further to "Applications in Manufacturing and Tooling". These typical applications are summarised in Fig. 7.1 and discussed in the following sections.

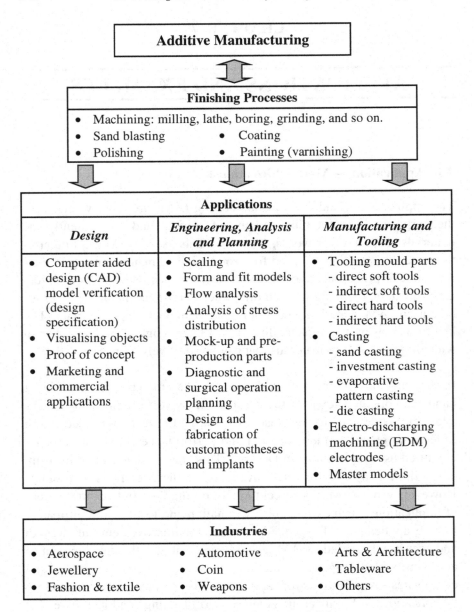

Figure 7.1: Typical application areas of AM.

The major advantage of AM technologies in manufacturing has been their ability to enhance and improve product development while at the same time reducing the costs and time required to take the product from concept to market. Sections 7.6–7.20 contain examples of applications in the aerospace, automotive, jewellery, coin, tableware, arts and architecture, building and construction, fashion and textile, weapons, music, marine and offshore, food and entertainment industries. These examples are by no means exhaustive, but they do represent their applications in a wide cross section of these industries.

7.2 Finishing Processes

As there are various influencing factors such as shrinkage, distortion, curling and accessible surface smoothness, it is necessary to apply some post-AM finishing processes to the parts just after they have been produced. These processes should be carried out before the AM parts are used in their desired applications. Furthermore, additional processes may be necessary in specific cases, such as when creating screw threads.

7.2.1 *Cutting processes*

In most cases, the resins or other materials used in AM systems can be subjected to conventional cutting processes, such as milling, boring, turning and grinding.

These processes are particularly useful in the following cases:

(1) Deviations in geometrical measurements or tolerances due to unpredictable shrinkage during the curing or bonding stages of the AM process.
(2) Incomplete generation of selected form features. This could be due to fine or complex-shaped features that were not achieved in AM.
(3) Clean removal of necessary support structures or other remaining material attached to the AM parts.

In all these cases, it is possible to achieve good surface finishing of the manufactured objects with numerical control (NC) and/or computer numerical control (CNC) machining.

7.2.2 Sandblasting and polishing

Sandblasting or abrasive jet deburring can be used as an additional cleaning operation or process to achieve higher surface quality. However, there is a trade-off in terms of accuracy. Should better finishing be required, additional polishing with super-fine abrasives or simple rotary sanding can also be used after sandblasting.

7.2.3 Coating

Appropriate surface coatings can be used to further improve the physical properties of the surface of plastic AM parts. One example is galvanic coating, a coating which provides very thin metallic layers on plastic AM parts.

7.2.4 Painting

Painting is applied fairly easily on AM parts made of plastics or paper. It is carried out mainly to improve the aesthetic appeal or for presentation purposes, such as for marketing or advertising presentations.

Once the AM parts are appropriately finished, they can then be used for the various areas of application as shown in Fig. 7.1.

7.3 Applications in Design

7.3.1 CAD model verification

This was the initial objective and strength of AM systems, in that designers often need the physical part to confirm the design that they have created in the CAD system. This is especially important for parts or products designed to fulfil aesthetic functions or are intricately designed to fulfil functional requirements.

7.3.2 *Visualising objects*

Designs created on CAD systems need to be communicated not only among designers within the same team, but also to other departments like the manufacturing and marketing teams. Thus, there is a need to create objects from the CAD designs for visualisation so that all these people will be referring to the same object in any communication. Tom Mueller in his paper entitled, "Application of Stereolithography in injection moulding"[1] characterises this necessity by saying:

> "Many people cannot visualise a part by looking at print. Even engineers and toolmakers who deal with print everyday require several minutes or even hours of studying a print. Unfortunately, many of the people who approve a design (typically senior management, marketing analysts, and customers) have much less ability to understand a design by looking at a drawing."

7.3.3 *Proof of concept*

Proof of concept relates to the adaptation of specific details to an object's environment or aesthetic aspects, like verifying that a car radio design is suitable and blends in well within a specific car. It also relates to adapting specific details of the design to its functional performance for a desired task or purpose, such as how the hinge of a laptop allows the opening of the laptop.

7.3.4 *Marketing and commercial applications*

Frequently, the marketing or commercial departments require a physical model for presentation and evaluation purposes, especially for assessment of the project as a whole. The mock-up or presentation model can even be used to produce promotional brochures and related materials for marketing and advertising even before the actual product becomes available.

7.4 Applications in Engineering, Analysis and Planning

Other than creating a physical model for visualisation or proofing purposes, designers are also interested in the engineering aspects of their designs. This invariably relates to the functions of the design. AM technologies become important as they are able to provide the information necessary to ensure sound engineering and function of the product. Furthermore, AM also helps to save development time and reduce costs. Based on the improved performance of processes and materials available in current AM technologies, some applications for functional models are presented in the following sections.

7.4.1 *Scaling*

AM technology allows easy scaling down (or up) of the size of a model by scaling the original CAD model. In a case of designing bottles for perfumes with different holding capacities, the designer can simply scale the CAD model appropriately for the desired capacities and view the renderings on the CAD software. With the selected or preferred capacities determined, the CAD data can be modified accordingly to create the corresponding AM model for visualisation and verification purposes (see Fig. 7.2).

Figure 7.2: Perfume bottles with different capacities.

7.4.2 *Form and fit*

Other than dealing with sizes and volumes, forms have to be considered from their aesthetics and functional standpoints as well. How a part fits into a design and into its environment are important aspects, which have to be addressed. For example, the wing mirror housing for a new car design has to be of a form that fits well with the general appearance of the exterior design. This will also include how it physically fits to the car door. The model will be used to evaluate how it satisfies both aesthetic and functional requirements.

Form and fit models are used not just in the automotive industries. They can also be used for industries involved in aerospace and others like consumer electronic products and appliances.

7.4.3 *Flow analysis*

Components that affect or are affected by air or fluid flow cannot be easily modified if produced by the traditional manufacturing routes. However, if the original 3D design data can be stored in a computer model, then any change of object data based on some specific tests can be realised with computer support. The flow dynamics of these products can also be computer simulated with software. Additionally, experiments with 3D physical models are frequently required to study product performance in air and liquid flow. Such models can be easily built using AM technology. Modifications in design can be done on computer and rebuilt for retesting very much faster than using traditional prototyping methods. Flow analyses are also useful for studying the inner sections of inlet manifolds, exhaust pipes, replacement heart valves[2] or similar products that at times can have rather complex internal geometries. Should it be required, transparent prototypes can also be additively manufactured to aid visualisation of internal flow dynamics. Typically, flow analyses are necessary for products manufactured in the aerospace, automotive, biomedical and shipbuilding industries.

7.4.4 *Stress analysis*

When conducting stress analyses using mechanical or photo-optical methods or otherwise, it is necessary to construct physical replicas of the part being analysed. If the material properties and features of the AM objects are similar to those of the actual functional parts, they can be used in these analytical methods to determine the stress distribution of the product.

An interesting consequence of the rising acceptance of AM is the use of topology optimisation for part design. This is a mathematical approach to achieve an optimised distribution of material in a given structure for a given load condition, for the ultimate purpose(s) of minimising material used, maximising load capabilities, maximising stiffness, and so on. This field was jump-started by Anthony Michell's famous 1904 paper on his Michell truss,[3] although topology optimisation only gained attention in recent years due to firstly the use of computers in solving algorithms and secondly the complexity of optimised parts, which necessitate the use of AM rather than conventional manufacturing and machining methods to produce the part. Topology optimisation can bring benefits to industries like the aviation industry,[4] where weight concerns and mechanical properties of parts are of high importance.

7.4.5 *Mock-up parts*

"Mock-up" parts, a term first introduced in the aircraft industry, are used for final testing of parts from different aspects. Generally, mock-up parts are assembled into the complete product and functionally tested at pre-determined conditions for fatigue and other tests. Some AM techniques are able to generate mock-ups very quickly to fulfil these functional tests before the design is finalised.

7.4.6 *Pre-production parts*

In cases where mass production will be introduced once the prototype design has been tested and confirmed, pilot-production runs of 10 or more parts are usual. The pilot-production parts are used to confirm tooling design and specifications. The necessary accessory equipment, such as fixtures, chucks, special tools and measurement devices required for the mass production process, are prepared and checked. Many of the AM methods available are able to quickly produce pilot-production parts, thus helping to shorten the process development time, thereby accelerating the overall time-to-market process.

7.4.7 *Diagnostic and surgical operation planning*

In combining engineering prototyping methodologies with surgical procedures, AM models can complement various imaging systems, such as ultrasonic imaging, magnetic resonance imaging (MRI) and computed tomography (CT) scanning, to produce anatomical models for diagnostic purposes. These AM models can also be used for surgical and reconstruction operation planning. This is especially useful in surgical procedures carried out by different teams of medical specialists, where interdepartmental communication is of essence. Surgeons can also familiarise themselves and practise with body parts before conducting the surgery. Several related examples and case studies can be found in Chap. 8.

7.4.8 *Design and fabrication of custom prostheses and*
implants

AM can be applied to design and fabricate customised prostheses and implants. A prosthesis or implant can be made from anatomical data inputs from imaging systems, such as MRI and CT. In cases like producing ear prostheses, a scan profile can be taken of the good ear to create a computer-mirrored exact replica replacement using AM technology. These models can be further refined and processed to create the actual prostheses or implants to be used directly on a patient. The ability to efficiently customise and produce such prostheses and implants is important, as standard sizes are not always an ideal fit for the patient. In fact, a less than ideal fit, especially for artificial joints and weight bearing implants, can often result in accumulative problems and damage to the surrounding tissue structures. More examples and case studies in these areas can be found in Chap. 8.

7.5 Applications in Manufacturing and Tooling

Central to the theme of rapid tooling is the ability to produce multiple copies of a prototype with functional material properties in short lead times. Apart from mechanical properties, materials can also include functionalities such as variations in colour, transparency, flexibility, and so on.

Rapid tooling can be classified as soft and hard tooling, and direct and indirect tooling,[5] as schematically shown in Fig. 7.3. Soft tooling, typically made of silicon rubber, epoxy resins, low melting point alloys and foundry sands, generally allows for only single casts or for small batch production runs. Hard tooling, on the other hand, is usually made from tool steels and generally allows for longer production runs.

Direct tooling is referred to when the tool or die is created directly by the AM process. To give an example in the case of injection moulding, the main cores, runner, gates and ejection systems can be produced directly using the AM process. On the other hand, in indirect tooling, only the

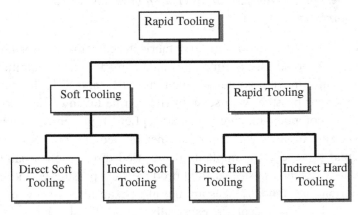

Figure 7.3: Classification of rapid tooling.

master pattern is created using the AM process. The mould, made of silicone rubber, epoxy resin, low melting point metal or ceramic, is then created from this master pattern.

7.5.1 *Direct soft tooling*

This is where the moulding tool is produced directly by the AM systems. Such tooling can be used for liquid metal sand casting, in which case the mould is destroyed after a single cast. Other examples, such as composite moulds, can be made directly using stereolithography (SLA). These are generally used in injection moulding of plastic components and can withstand between 100 and 1000 shots. As these moulding tools can typically only support a single cast or small batch production before breaking down, they are classified as soft tooling. The following section lists several examples of direct soft tooling methods.

7.5.1.1 *Selective laser sintering of sand casting moulds*

Sand casting moulds can be produced directly using the selective laser sintering (SLS) process. Individual sand grains are coated with a polymeric binder. Laser energy is applied to melt this binder, thereby coating the individual sand grains and bonding the grains of sand together in the shape of a mould.[6] Accuracy and surface finish of the metal castings produced from such moulds are similar to those produced by conventional sand casting methods. Functional prototypes can be produced this way, and if modifications are necessary, a new prototype can be produced within a few days.

7.5.1.2 *Direct AIM (accurate clear epoxy solid injection moulding)*

A rapid tooling method developed by 3D computer-aided design and computer-aided manufacturing (CAD/CAM) systems uses the SLA to produce resin moulds that allow direct injection of thermoplastic materials. Known as the Direct AIM (accurate clear epoxy solid (ACES) injection moulding),[7] this method is able to produce high levels of accuracy. However, build times using this method is relatively slow on standard SLA machines. Also, because the mechanical properties of these moulds are generally poor, tool damage can occur during the ejection of the part. This is more evident when producing geometrically complex parts using these moulds.

7.5.1.3 *SLA composite tooling*

This method builds moulds with thin shells of resin with the required surface geometry, which is then backed-up with aluminium powder-filled epoxy resin to form the rest of the mould tooling.[8] This method is advantageous in that higher mould strengths can be achieved when compared to those produced by the Direct AIM method, which builds a solid SLA resin mould. To further improve the thermal conductivity of the mould, aluminium shot can be added to back the thin shell, thus promoting faster build times for the mould tooling. Other advantages of this method include higher thermal conductivity of the mould and lower tool development costs when compared to moulds produced by the Direct AIM method.

7.5.2 *Indirect soft tooling*

In this rapid tooling method, a master pattern is first produced using AM. From the master pattern, a mould tooling can be built out of an array of materials such as silicone rubber, epoxy resin, low melting point metals and ceramics.

7.5.2.1 *Arc spray metal tooling*

Using metal spraying on the AM model, it is possible to create an injection mould very quickly that can be used to mould a limited number of prototype parts. A typical metal spray process for creating injection mould is shown in Fig. 7.4. The metal spraying process is operated manually with a handheld gun. An electric arc is introduced between two wires, which melt the wires into tiny droplets.[9] Compressed air blows out the droplets in small layers of approximately 0.5 mm of metal.

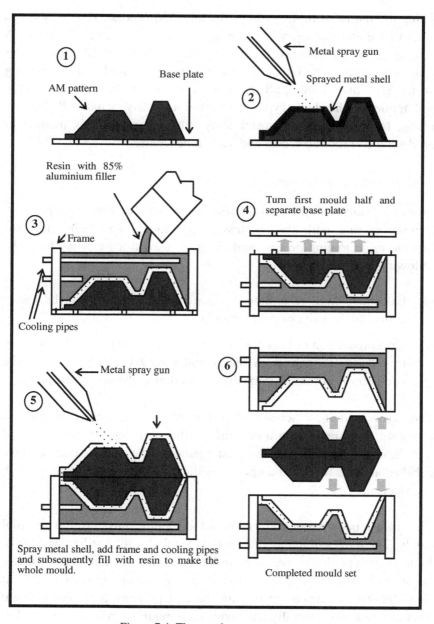

Figure 7.4: The metal arc spray system.

Using the master pattern produced by any AM process, this is mounted onto a base and bolster, which are then layered with a release agent. A coating of metal particles using the arc spray is then applied to the master pattern to produce the female form cavity of the desired tool. Depending on the type of tooling application, a reinforcement backing is selected and applied to the shell. Types of backing materials include filled epoxy resins, low melting point metal alloys and ceramics. This method of producing soft tooling is cost and leads time saving.

7.5.2.2 *Silicone rubber moulds*

In manufacturing functional plastic, metal and ceramic components, vacuum casting with the silicone rubber mould has been the most flexible rapid tooling process and the most used to date. They have the following advantages:

(1) Extremely high resolution of master model details can be easily copied to the silicone cavity mould.

(2) Gross reduction of backdraft problems (i.e., die lock, or the inability to release the part from the mould cavity because some of the geometry is not in the same draw direction as the rest of the part).

The master pattern, attached with a system of sprue, runner, gating and air vents, is suspended in a container. Silicone rubber slurry is poured into the container engulfing the master pattern. The silicone rubber slurry is baked at 70°C for 3 h and upon solidification, a parting line is cut with a scalpel.

The master pattern is removed from the mould, thus forming the tool cavity. The halves of the mould are then firmly taped together. Materials, such as polyurethane, are poured into the silicone tool cavity under vacuum to avoid asperities caused by entrapped air. Further baking at 70°C for 4 h is carried out to cure the cast polymer part. The vacuum casting process is generally used with such moulds. Each silicone rubber mould can produce up to 20 polyurethane parts before it begins to break apart.[10] These problems are commonly encountered when using hard moulds, making it necessary to have expensive inserts and slides. They can be cumbersome and take a longer time to produce. These are virtually eliminated when the silicone moulding process is used.

AM models can be used as master patterns for creating these silicone rubber moulds. Fig. 7.5(a)–(f) describes the typical process of creating a silicone rubber mould and the subsequent urethane-based part.

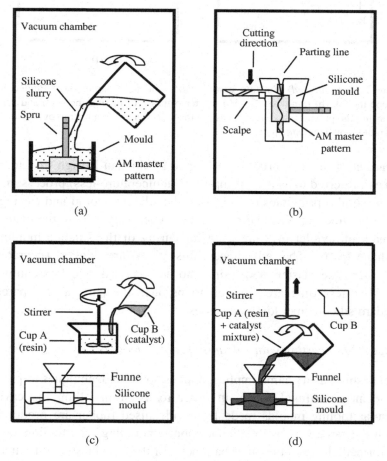

Figure 7.5: Vacuum casting with silicone moulding. (a) Producing the silicone mould, (b) removing the AM master pattern, (c) mixing the resin and catalyst, (d) casting the polymer mixture, (e) cast urethane part cured in a baking oven and (f) the final rapid tooled urethane part.

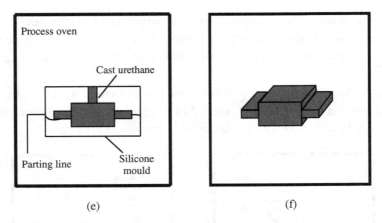

(e)   (f)

Figure 7.5: (*Continued*) Vacuum casting with silicone moulding. (a) Producing the silicone mould, (b) removing the AM master pattern, (c) mixing the resin and catalyst, (d) casting the polymer mixture, (e) cast urethane part cured in a baking oven and (f) the final rapid tooled urethane part.

A variant of this is a process developed by Shonan Design Co Ltd. This process, referred to as the "Temp-less" (temperature-less) process, makes use of similar principles of preparing the silicone mould and casting the liquid polymer, except that no baking is necessary to cure the materials. Instead, ultraviolet rays are used for curing of the silicone mould and urethane parts. The advantages this gives are higher accuracy of replicating the master model since no heat is used, and less equipment and 70% less time are required to produce the parts as compared to standard silicone moulding processes.[11]

7.5.2.3 *Spin casting with vulcanised rubber moulds*

Spin casting, as its name implies, applies spinning techniques to produce sufficient centrifugal forces in order to assist in filling the cavities. Circular tooling moulds made from vulcanised rubber are produced in much the same way as in silicone rubber moulding. The tooling cavities are formed closer to the outer parameter of the circular mould to increase centrifugal forces. Polyurethane or zinc-based alloys can be cast using this method.[12] This process is particularly suitable for producing low volumes of small zinc prototypes that will ultimately be mass-produced by die-casting.

7.5.2.4 *Castable resin moulds*

Similar to the silicone rubber moulds, the master pattern is placed in a mould box with the parting line marked out in plasticine.[13] The resin is painted or poured over the master pattern until there is sufficient material for one half of the mould. Different tooling resins may be blended with aluminium powder or pellets so as to provide different mechanical and thermal properties. Such tools are able to typically withstand up to 100–200 injection moulding shots.

7.5.2.5 *Castable ceramic moulds*

Ceramic materials that are primarily sand based can be poured over a master pattern to create the mould.[14] The binder systems can vary with different preferences of binding properties. For example, in colloidal silicate binders, the water content in the system can be altered to improve shrinkage and castability properties. The ceramic–binder mix can be poured under vacuum conditions and vibrated to improve the packing of the material around the master pattern.

7.5.2.6 *Plaster moulds*

Casting into plaster moulds has been used to produce functional prototypes.[15] A silicone rubber mould is first created from the master pattern and a plaster mould is then made from this. Molten metal is then poured into the plaster mould, which is broken away once the metal has solidified. Silicone rubber is used as an intermediate stage because the pattern can be easily separated from the plaster mould.

7.5.2.7 *Casting*

In the metal casting process, a metal, usually an alloy, is heated until it is in a molten state, whereupon it is poured into a mould or die that contains a cavity. The cavity will contain the shape of the component or casting to be produced. Although there are numerous casting techniques available, three main processes are discussed here: (1) the conventional sand casting, (2) investment casting and (3) evaporative casting processes. AM models render themselves well to be the master patterns for the creation of these metal dies.

Sand casting moulds are similarly created using AM master patterns. AM patterns are first created and placed appropriately in the sand box. Casting sand is then poured and packed very compactly over the pattern. The box (cope and drag) is then separated and the pattern carefully removed leaving behind the cavity. The box is assembled together again and molten metal is cast into the sand mould. Sand casting is the cheapest and most practical for casting large parts. Fig. 7.6 shows a cast metal mould resulting from an AM pattern made by laminated object manufacturing (LOM).

Figure 7.6: Cast metal (left) and AM pattern for sand casting.
Courtesy of Helisys, Inc.

Another casting method, the investment casting process, is probably the most important moulding process for casting metal. Investment casting moulds can be made from AM master patterns. The pattern is usually wax, foam, paper or other materials that can be easily melted or vaporised. The pattern is dipped in slurry ceramic compounds to form a relatively strong coating, or investment shell, over it.[16] This is repeated until the shell builds up thickness and strength. The shell is then used for casting, with the pattern being melted away or burned out of the shell, resulting in a ceramic cavity. Molten metal can then be poured into the mould to form the object. The shell is then cracked open to release the desired object in the mould. The investment casting process is ideal for casting miniature parts with thin sections and complex features. Fig. 7.7 shows schematically the investment casting process from an AM-produced wax master pattern, whereas Fig. 7.8 shows an investment casting mould resulting from an AM pattern.

The third casting process is the evaporative pattern casting. As its names implies, this uses an evaporative pattern, such as polystyrene foam, as the

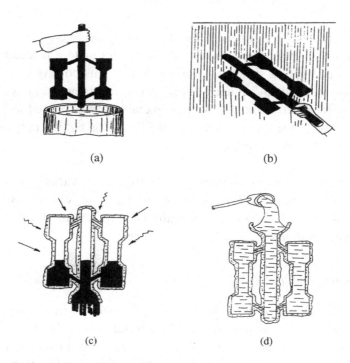

(a)   (b)

(c)   (d)

Figure 7.7: Schematic diagram of the shell investment casting process. (a) Pattern clusters are dipped in ceramic slurry, (b) refractory grain is sifted onto the coated patterns. Steps (a) and (b) are repeated several times to obtain desired shell thickness, (c) after the mould material has set and dried, the patterns are melted out of the mould and (d) hot moulds are filled with metal by gravity, pressure vacuum or centrifugal force.

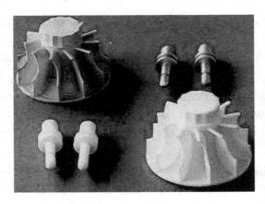

Figure 7.8: Investment casting of an impeller from an AM pattern.

master pattern. This pattern can be produced using the SLS process along with CastForm™ polystyrene material. The master pattern is attached to sprue, riser and gating systems to form a "tree". This polystyrene "tree" is then surrounded by foundry sand in a container and vacuum compacted to form a mould. Molten metal is then poured into the container through the sprue. As the metal fills the cavity, the polystyrene evaporates with very low ash content.[17]

The part is cooled before the casting is removed. A variety of metals, such as titanium, steel, aluminium, magnesium and zinc can be cast using this method. Fig. 7.9 shows schematically how an AM master pattern is used with the evaporative pattern casting process.

7.5.3 *Direct hard tooling*

Hard tooling produced by AM systems has been a major topic for research in recent years. Although several methods have been demonstrated, much research is still being carried out in this area. The advantages of hard tooling produced by AM methods are fast turnaround times to create highly complex-shaped mould tooling for high volume production. Fast response to modifications in generic designs can be almost immediate. The following are some examples of direct hard tooling methods.

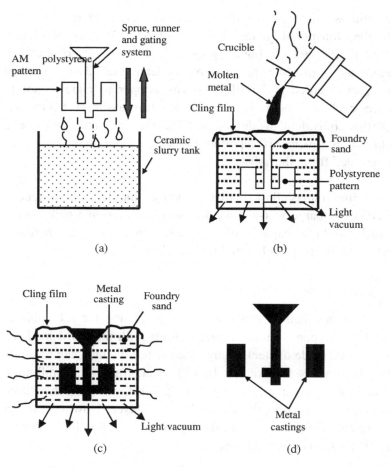

Figure 7.9: Evaporative pattern casting process. (a) Polystyrene AM pattern "tree" is coated by dipping with a ceramic slurry and air-dried, (b) coated AM pattern is packed with foundry sand in a container. The container is sealed with cling film and vacuumed to compact the sand further, (c) the polystyrene pattern evaporates as the molten metal is cast into the mould. The casting is then left to cool and (d) after solidification, the final cast parts are broken away from the sprue, runner and gating system.

7.5.3.1 *RapidTool*^{TM}

RapidTool^{TM} is a technology previously patented by DTM Corporation to produce metal moulds for plastic injection moulding directly from the SLS Sinterstation^{TM}. The moulds are capable of being used in the conventional injection moulding machines to mould the final product with the functional material.[18] The CAD data is fed into the Sinterstation™, which bonds polymeric binder coated metal beads together using the SLS process. Next, debinding takes place and the green part is cured and infiltrated with copper to make it solid. The furnace cycle is about 40 h with the finished part having similar properties equivalent to aluminium. The finished mould can be easily machined. Shrinkage is reported to be no more than 2%, which is compensated for in the software.

Typical time frames allow relatively complex moulds to be produced in 2 weeks as compared to 6–12 weeks using conventional techniques. The finished mould is capable of producing up to tens of thousands of injection-moulded parts before breaking down.

7.5.3.2 *Laminated metal tooling*

This is another method that may prove promising for RT applications. The process applies laminated metal sheets with the LOM method. The sheets can be made of steel or any other material which can be cut by the appropriate means, for example by CO_2 laser, water jet or milling, based on the LOM principle.[19] The CAD 3D data provides the sliced 2D information for cutting the sheets layer-by-layer. However, instead of bonding each layer as it is cut, the layers are all assembled after cutting and either bolted or bonded together.

7.5.3.3 *Selective laser melting tooling*

The selective laser melting (SLM) technology was developed by SLM Solutions GmbH, formerly known as MTT Technologies GmbH. The process uses a very high-powered infrared laser to melt metal powders directly. The powders available for use by this technology include stainless steel tool steel, titanium, aluminium, cobalt-chrome, and so on.[20] The process is capable of producing parts of 100% density, so that further infiltration is not required which is the case for sintering. Hence, the resulting part has high strength and dimensional accuracy.

7.5.3.4 *ProMetal™ rapid tooling*

Based on MIT's three-dimensional printing (3DP) process, the ProMetal™ Rapid Tooling System is capable of creating steel parts for the tooling of plastic injection moulding parts, lost foam patterns and vacuum forming. This technology uses an electrostatic ink jet print head to eject liquid binders onto the powder, selectively hardening slices of an object a layer at a time. A fresh coat of metal powder is spread on top and the process repeats until the part is complete. Loose powders act as supports for the object to be built. The AM part is then infiltrated at furnace temperatures with a secondary metal to achieve full density. Toolings produced by this technology for use in injection moulding have reported withstanding pressures up to 30,000 psi (200 MPa) and survived 100,000 shots of glass-filled nylon.[21]

7.5.4 *Indirect hard tooling*

There are numerous indirect AM tooling methods that fall under this category and this number continues to grow. However, many of these processes remain largely similar in nature except for small differences such as binder system formulations or type of system used. Processes include the rapid solidification process (RSP), Ford's (UK) Sprayform, Cast Kirksite Tooling, CEMCOM's chemically bonded ceramics (CBC) and Swift Technologies Ltd. "SwiftTool", just to name a few. This section will only cover selected processes that can also be said to generalise all the other methods under this category. In general, indirect methods for producing hard tools for plastic injection moulding generally make use of casting of liquid metals or steel powders in a binder system. For the latter, debinding, sintering and infiltration with a secondary material are usually carried out as post-processes.

7.5.4.1 *3D Keltool*

The 3D Keltool process has been developed by 3D Systems to produce a mould in fused powdered steel.[22] The process uses an SLA model of the tool for the final part that is finished to a high quality by sanding and polishing. The model is placed in a container where silicone rubber is poured around it to make a soft silicone rubber mould that replicates the female cavity of the SLA model. The mould is then placed in a box and silicone rubber is poured around it to produce a replica copy of the SLA model. This silicone rubber is then placed in a box and a proprietary mixture of metal particles, such as tool steel, and a binder material is poured around it, cured and separated from the silicone rubber model. This is then fired to eliminate the binder and sinter the metal particles together. The sintered part is about 70% steel. The 30% void is then infiltrated with copper to give a solid mould, which can be used in injection moulding.

An alternative to this process is described as the reverse generation process. This uses a positive SLA master pattern of the mould and requires one fewer step. This process claims that the CAD solid model to injection-moulded production part can be completed in 4–6 weeks. Cost savings of around 25–40% can be achieved when compared to that of conventional machined steel tools.

7.5.4.2   *Electro-discharge machining electrodes*

A method successfully tested in research laboratories but so far not widely applied in the industry is the possible manufacturing of copper electrodes for EDM (electro-discharge machining) processes using AM technology. To create the electrode, the AM-created part is used to create a master for the electrode. An abrading die is created from the master by making a cast using an epoxy resin with an abrasive component. The resulting die is then used to abrade the electrode. A specific advantage of the SLS procedure (see Section 5.1) is the possible usage of other materials. Using copper in the SLS or SLM process, it is possible to generate the electrodes used in EDM quickly and affordably.

7.5.4.3   *Ecotool*

This is a development between the Danish Technological Institute (DTI) in Copenhagen, Denmark, and the TNO Institute of Industrial

Technology of Delft in Holland. The process uses a new type of powder material with a binder system to rapidly produce tools from AM models. As its name implies, the binder is environmentally friendly in that it uses a water-soluble base. An AM master pattern is used and a parting line block produced. The metal powder–binder mixture is then poured over the pattern and parting block and left to cure for an hour at room temperature. The process is repeated to produce the second half of the mould in the same way. The pattern is then removed and the mould is baked in a microwave oven.

7.5.4.4 *Copy milling*

Although not broadly applied nowadays, AM master patterns can be provided by manufacturers to their vendors for use in copy milling, especially if the vendor for the required parts is small and does not have the more expensive but accurate CNC machines. In addition, the principle of generating master models only when necessary allows some storage space to be saved. The limitation of this process is that only simple geometrical shapes can be made.

7.6 Aerospace Industry

The aerospace industry is the leading pioneer industry that other industries look to for a glimpse of what is new on the horizon. In particular, the civil aviation industry is known for having strict requirements for its products due to passenger safety concerns. This industry has incorporated AM throughout its product development lifecycle — from the design concept to maintenance and repairs. With the various advantages that AM technologies promise, it is only natural that the aerospace industry, encompassing both space exploration as well as military and civil aviation, continues to discover and find new applications and invest in research to make them possible. The following are a few examples.

7.6.1 *Unmanned aerial vehicles*

The production of unmanned aerial vehicles (UAVs) is highly complex, low-volume, and requires rapid design iterations. Particularly with the absence of passenger safety regulations, UAVs and the unmanned aerial systems (UAS) industry have a lot to benefit from AM. AM allows designers to skip the fabrication of tools and go straight to the finished parts. Proven to be effective in creating light-weight, durable and complex aerospace parts, AM has been very quickly adopted by UAV manufacturing companies.

SelectTech Geospatial's Advanced Manufacturing Facility used a Dimension 3D printer from Stratasys to build and fly the world's first 3D printed UAS that can take off and land on its own landing gear.[23] Taking a trial-and-error approach, SelectTech refined the entire airframe through discoveries and corrections from iterative 3D-printed physical prototypes. Eliminating the need for expensive tooling during the design process allowed SelectTech designers to focus on creating lighter weight structures out of ABS material, boosting efficiencies, strengthening parts and speeding up the overall development process. As compared to 2D laser cutting, lead times fell from 2 months to 2 weeks, and costs fell from $30,000 to $5150. The success of this project demonstrated that complicated and sophisticated products can be made, tested and manufactured through a rapid-response model using AM without committing to a final design and manufacturing tools. The completed design, with a 1.2-m wingspan, is shown in Fig. 7.10.

Aurora Flight Sciences is a manufacturer of advanced UAVs and an aerospace contractor. The company successfully created and flew the world's first 3D printed jet-powered aircraft using Stratasys' Fortus machines.[24] 26 out of a total of 34 components were 3D printed. For instance, the exhaust duct cover was printed with ULTEM 1010 thermoplastic, whereas the thrust vectoring duct was manufactured with DMLS of Inconel 718. The aircraft has a 2.9-m wingspan and can cruise at 60–70 knots while carrying a 0.2-kg camera. AM solves the problems of design constraints that traditional fabrication techniques often pose, and in this case has halved the build time by obviating tooling.

Figure 7.10: SelectTech Geospatial used a Stratasys Dimension 3D printer to additively manufacture this UAS.
Courtesy of Stratasys and SelectTech Geospatial

7.6.2 *Flight instruments*

The civil aviation industry has high performance and safety standards to comply with, and has now accepted AM in the manufacturing of parts such as air grates, panel covers, heating, ventilation, and air conditioning (HVAC) ducts, power distribution panes, some interior parts as well as various mounting and attachment hardware. AM technology has also been applied to manufacturing flight instruments.

Kelly Manufacturing Company, founded in 1932, manufactures the popular R.C. Allen line of aircraft instrumentation, such as the M3500 "turn-and-slip" indicators. By contracting Rapid Processing Solutions Inc., they are using Stratasys' Fortus machines to manufacture their M3500 indicator toroid housings from ULTEM 9085. Through the implementation of Fused Deposition Modeling (FDM) they are now able to produce 500 pieces of toroid housings in one overnight run as opposed to their traditional method of using a urethane-moulded technique which took 4 weeks.[25] In addition, the use of FDM has helped them reduce per-piece cost by 5%, and has completely eliminated tooling costs and removed the need to hand-sand the final product.

7.6.3 *Jet engines and parts*

GE Aviation, part of the world's biggest manufacturing group, has made significant investments in AM to build more than 85,000 fuel nozzles for its new LEAP jet engines as part of a joint venture with Snecma S.A. of France. The new LEAP engines will power the new Airbus A320neo, Boeing 737 MAX and Comac C919 jetliners.[26] Previously, the fuel nozzles were assembled from 20 different parts in a labour-intensive and wasteful process.

Instead, GE Aviation will use EOS GmbH's DMLS systems to manufacture complex and intricately-designed cobalt-chromium parts with five times fewer brazes and welds, ultimately gaining a five-fold improvement in durability.[27] Aside from the cost and time-saving benefits of using AM, the 3D-printed fuel nozzles will each be 25% lighter and five times more efficient than conventionally manufactured nozzles. Each LEAP engine will have nineteen 3D-printed fuel nozzles. A picture of one of these fuel nozzles is shown in Fig. 7.11.

Figure 7.11: GE Aviation's 3D printed fuel nozzle to be used in the latest LEAP engine. Courtesy of GE Aviation

7.6.4 *Space exploration and zero-G 3D printing*

When it comes to building highly customised vehicles and testing them in punishing environments, the National Aeronautics and Space Administration (NASA) knew that only AM offered the design flexibility and quick turnaround process. The NASA team used 3D printing to build a Mars rover with about 70 parts built digitally and directly from computer designs in the heated chamber of a Stratasys 3D printer.[28] Using FDM, the process created complex shapes such as an ear-shaped exterior housing, which is so deep and twisted that it would have been economically unfeasible to build otherwise. NASA engineers used AM to build prototypes to test the form, fit and function of parts that they will eventually make in other materials. 3D printing allowed designers to solve difficult challenges before committing to expensive tooling, ensuring machined parts are based on the best possible design.

NASA has also tested 3D printing in microgravity environments, rather than the 1-G environments experienced on earth. 3D printing aboard space stations offers a fast and cheap way to manufacture parts on demand, reducing the need for costly spares on long-term space missions with weight restrictions. The proof-of-concept 3D Printing In Zero-G Technology Demonstration occurred aboard the International Space Station and focused on relatively low-temperature ABS thermoplastic raw material.[29] The demonstration showed that 3D printers work normally in space. This demonstration lays the groundwork for establishing an on-demand machine shop in space, hopefully with plastic and metal AM production capabilities.

It is clear that there are still a lot of untapped benefits that AM can provide for the aerospace industry. Many corporations and research laboratories are funding new projects into using AM for the production of parts and planes. Airbus has done considerable work in the 3D printing of polymer and metal parts, and plans to 3D print 30 tons of metal parts on 40 AM machines monthly by December 2018.[30] UTC Aerospace Systems, formed as a merger between Hamilton Sundstrand and Goodrich, has opened a new 1800-m^2 Materials and Process

Engineering laboratory as an $8-million investment, and established a Materials Engineering Center of Excellence at University of Connecticut as part of a 5-year $1-million commitment to AM research.[31] Although there are still significant challenges before AM can be a fully practical alternative to traditional production methods, AM has already made its marks on designing, testing, tooling and production in the aerospace industry. The versatility of this technology will extend beyond aerospace applications.

7.7   Automotive Industry

As early adopters of rapid prototyping (an earlier name for AM), the automotive industry has more 3D printing applications than any other industry. The automotive industry has always been a technology- and innovation-driven industry, where consumers' demands for design, safety, comfort and environmental friendliness create various challenges for companies to overcome.

7.7.1 *Multi-material manufacturing and prototyping*

Bentley Motors Ltd. utilised Stratasys' PolyJet technology to speed up the production of small-scale models and full-size parts for assessment and testing before production on the assembly line. Every part of the car is first prototyped in miniature. The accuracy of the Objet30 Pro 3D printer enabled the operations and projects team to emulate a full-size part and scale it down to a one-tenth scale model.[32] Once the miniature model is approved, selected full-sized parts and multi-material parts can be produced without assembly. One such miniature model is shown in Fig. 7.12.

Stratasys' Objet500 Connex 3D printer has given the design studio team the power to combine a variety of materials. A single prototype can combine rigid and rubber-like, and clear and opaque materials with no assembly required, such as a rubber tyre on a plastic wheel rim. Polyjet's rubber-like material also enables Bentley to simulate rubber according to different levels of hardness and tear resistance.

Figure 7.12: A 3D printed multi-material miniature model of a Bentley vehicle.
Courtesy of Stratasys and Bentley Motors

Using traditional methods, such a prototype would have been too costly and time consuming to build and it might not have been always possible to include all the fine details of the design. Fabrication of the model based on drawings was often subject to human interpretation and was therefore error prone, thus further complicating the prototyping process. All these difficulties were avoided by using AM technology, as the production of the model was based entirely on the CAD model that was created before it was 3D printed.

AM technology has shrunk the prototype development time from months to 2 weeks, and considerable time and cost savings were achieved. The ability to 3D print parts with different materials together has also revolutionised the entire Bentley Motors design process.

7.7.2 Making cars with ergonomic hand-held tools

Although AM has become an integral part of product development in automotive manufacturing, German automaker BMW is extending FDM's application to other areas, such as building ergonomically designed assembly line hand tools that perform better than conventional tools. These tools, manufactured by Stratasys' 3D Production System, are easier and more comfortable to use for repetitive processes, which ultimately improves productivity. For example, one hand-held device's weight was reduced by 72% with a sparse-fill build technique, taking 1.3 kg off the device. The FDM process has turned out to be a cost saving alternative to conventional manufacturing methods, with cost reductions in engineering documentation, warehousing and manufacturing.[33] BMW has also come up with a simple flow chart to determine if FDM should be used. The criteria are temperature, chemical exposure, precision and mechanical load. The layered FDM manufacturing process is also suitable for the production of complex bodies compared with conventional metal-cutting processes, which would be difficult and expensive to produce.

FDM enables designers to create devices that utilise the advantages of AM. This helps enterprises with their product development and offers an alternative method for manufacturing components in small numbers.

7.7.3 *Selling collector models, physical and virtual*

Ford has been at the forefront of AM for 25 years and was involved with the invention of 3D printing in the 1980s. Using various systems such as SLS, FDM and SLA, Ford has used and continues to use AM to speed up the production of prototype parts and individual components.

More interestingly, Ford has also come up with an innovative method to utilise AM to help sell its cars. Ford sells both printed 3D models of its own cars, as well as .STL files for customers, hobbyists and model collectors to buy and print by themselves.[34] Although the pre-printed models measure around 10 cm long, customers can buy the cheaper .STL file instead and scale the model as they desire.

7.7.4 *World's first 3D printed car*

At the 2014 International Manufacturing Technology Show in Chicago, the world's first 3D printed car, called the Strati, was printed in one piece in 44 h.[35] The entire car is made of thermoplastic, except for the mechanical components such as the battery, motor and suspension which had to be separately assembled. Made by Local Motors in collaboration with Oak Ridge National Laboratory, the prototype has developed into the LM3D, which at the time of writing will soon be launched and made available for retail to the public.[36] Local Motors hopes to eventually make about 90% of the LM3D using 3D printing, and to offer the car with different, highly customisable aesthetic features for each individual user.

Previous attempts have been made to 3D print individual car parts. KOR EcoLogic Inc., a design and engineering consultancy firm based in Winnipeg, Canada, is responsible for the world's first prototype car, called URBEE, which had its entire body 3D printed.[37] Before printing the full-sized body panels, the computer files were first verified to be correct by printing a 1/6th scale model, as shown in Fig. 7.13.

Figure 7.13: The 1/6th scale model of URBEE.
Courtesy of Dana McFarlane and KOR Ecologic

KOR EcoLogic worked closely with Stratasys to print all the vehicle's exterior components. The body was produced in ABS plastic using FDM over a continuous 2500-h period. Unlike traditional cars made from sheet metal, the URBEE car framework did not require connecters, nuts and bolts and was constructed out of large single pieces of plastic. The entire finished 3D printed product is shown in Fig. 7.14. The company has moved on to the next phase of producing URBEE 2.

Figure 7.14: The URBEE's entire body was 3D printed.
Courtesy of KOR EcoLogic

3D printing is still in its early stage for manufacturing and has not yet fully replaced the automotive industry's current high speed and high volume direct production process. However, the technology is definitely making its impact in the product development and testing stages.

7.8 Jewellery Industry

The jewellery industry has traditionally been regarded as heavily craft based, and automation is generally restricted to the use of machines in the various individual stages of jewellery manufacturing. The use of AM technology in jewellery design and manufacture offers a significant breakthrough in this industry and has been subject to research since the 1990s. In an experimental computer-aided jewellery design and manufacturing system jointly developed by Nanyang Technological University (NTU) and Gintic Institute of Manufacturing Technology in Singapore, the SLA from 3D Systems was used successfully to create fine jewellery models.[38] These were used as master patterns to create the rubber moulds for making wax patterns. The wax patterns were later used in investment casting of precious metals as the end products, as shown in Fig. 7.15. In an experiment with the design of rings, the overall quality of the SLA models were found to be promising, especially in the generation of intricate details in the design. However, due to the nature of the step-wise building of the model, steps at the "gentler" slope of the model were visible. With the use of better resin and finer layer thickness, this problem was reduced but not fully eliminated. Further processing was found to be necessary, and abrasive jet deburring was identified to be most suitable.[39]

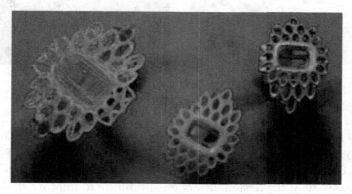

Figure 7.15: A two-times scaled SLA model to aid visualisation (left), full-scale wax pattern produced from the silicone rubber moulding (centre) and an investment cast silver alloy prototype of a brooch (right).

Today, there exist commercial turnkey AM systems designed specifically for jewellery makers, the most prominent of which is the Solidscape 3Z series. The 3Z series makes wax patterns for investment casting and has a small build size and high resolution and accuracy. Jenny Wu, a Los Angeles-based architect and designer, pioneered her LACE series of jewellery using Solidscape 3D printers, specifically the Solidscape MAX[2]. Jenny Wu says that "it's been fun to look at pieces as a whole — much more rewarding than seeing an image of a ring, for example". She has designed bracelets, earrings, necklaces, rings and other accessories. According to Solidscape, there are approximately 5500 3D printers being used in the jewellery industry worldwide.[40] Two LACE Papilio Rings made from polished sterling silver and 14-K plated rose gold are shown in Fig. 7.16.

Figure 7.16: LACE Papilio Rings.
Courtesy of LACE by Jenny Wu

Though post-processing of 3D printed patterns or casted jewellery pieces is likely necessary in the manufacturing process, the ability to create models quickly in a few hours instead of days or weeks, depending on the complexity of the design, offers great promise to improve design and manufacture in the jewellery industry.

7.9 Coin Industry

Similar to the jewellery industry, the mint industry has traditionally been regarded as very labour intensive and craft based. It relies primarily on the skills of trained craftsmen in generating "embossed" or relief designs on coins and other related products. In another experimental coin manufacturing system using CAD/CAM, CNC and AM technologies developed by NTU and Gintic Institute of Manufacturing Technology in Singapore, the SLA from 3D Systems was used successfully with Relief Creation Software to create tools for coin manufacture.[41] In the system involving AM technology, its working methodology consists of several steps.

First, 2D artwork is read into ArtCAM, the CAD/CAM system used in the system, using a Sharp JX A4 scanner. Fig. 7.17 shows the 2D artwork of a series of Chinese characters and a roaring dragon. In the ArtCAM software, the scanned image is reduced from a colour image to a monochrome image with the fully automatic "Grey Scale" function. Alternatively, the number of colours in the image can be reduced using the "Reduce Colour" function. A colour palette is provided for colour selection and the various areas of the images are coloured, using either different sizes and types of brushes or the automatic flood fill function.

Figure 7.17: Two-dimensional artwork of a series of Chinese characters and a dragon.

The second step is the generation of surfaces. The shape of a coin is generated to the required size in the CAD system for model building. A triangular mesh file is produced automatically from the 3D model. This is used as a base onto which the relief data is wrapped and later combined with the relief model to form the finished part.

The third step is the generation of the relief. In creating the 3D relief, each colour in the image is assigned a shape profile. There are various fields that control the shape profile of the selected coloured region, namely the overall general shape for the region, the curvatures of the profile (convex or concave), and the maximum height, base height, angle and scale. The relief detail generated can be examined in a dynamic Graphic Window within the ArtCAM software. Fig. 7.18 illustrates the 3D relief of the dragon artwork.

Figure 7.18: Three-dimensional relief of artwork of the roaring dragon.

The fourth step is the wrapping of the 3D relief onto the coin surface. This is done by wrapping the three-dimensional relief onto the triangular mesh file generated from the coin surfaces. This is a true surface wrap and not a simple projection. The wrapped relief is also converted into triangular mesh files. The triangular mesh files can be used to produce a 3D model suitable for colour shading and machining. The two sets of triangular mesh files, of the relief and the coin shape, are automatically combined. The resulting model file can be colour-shaded and used by the SLA to build the prototype.

The fifth step is to convert the triangular mesh files into the .STL file format, which would be used to build the AM model. After the conversion, the .STL file is sent to the SLA to create the 3D coin pattern, which will be used for proofing of design.

7.10 Tableware Industry

In another application to a traditional industry, the tableware industry, CAD and AM technologies are used in an integrated system to create better designs in a faster and more accurate manner. The general methodology used is similar to that used in the jewellery and coin industries. Additional computer tools with special programs developed to adapt decorative patterns to different variations of size and shape of tableware are needed for this particular industry.[42] Also, a method for generating motifs along a circular arc has been developed to supplement the capability of such a system.[43]

The general steps involved in the art to part process for the tableware include the following:

(1) Scanning of the 2D artwork.
(2) Generation of surfaces.
(3) Generation of 3D decoration reliefs.
(4) Wrapping of reliefs on surfaces.
(5) Converting triangular mesh files to STL file.
(6) Building of model by the AM system.

Two AM systems are selected for experimentation in the tableware system. One is 3D Systems' SLA, and the other is Cubic Technologies' LOM. The SLA has the advantages of being a pioneer and a proven technology with many excellent case studies available. It is also advantageous to use in tableware design as the material is translucent and thus allows designers to view the internal structure and details of tableware items like tea pots and gravy bowls. On the other hand, the use of LOM has its own distinct advantages. Its material cost is much lower and because it does not need support in its process (unlike the SLA), it saves a lot of time in both pre-processing (deciding where and what supports to use) and post-processing (removing the supports). Examples of dinner plates produced using the systems are shown in Fig. 7.19.

Figure 7.19: Dinner plate prototype built using SLA (left) and LOM (right).

In an evaluation test of making the dinner plate prototype, it was found that the LOM prototype was able to recreate the floral details more accurately. The dimensional accuracy was slightly better in the LOM prototype. In terms of build time, including pre- and post-processing, the SLA was about 20% faster than the LOM process. However, with sanding and varnishing, the LOM prototype appeared to be a better model which could be used afterwards to create the plaster of Paris moulds for the moulding of ceramic tableware (see Fig. 7.20 for a tea pot built using LOM). Apart from these technical issues, the initial investment, operating and maintenance costs of the SLA were considerably higher than that of the LOM, estimated to be about 50–100% more.

Figure 7.20: LOM model of a tea pot.
Courtesy of Champion Machine Tools, Singapore

In the ceramic tableware production process, the LOM model can be used directly as a master pattern to produce the block mould. The mould is made of plaster of Paris. The result of this trial is shown in Fig. 7.21. The trials highlighted the fact that plaster of Paris is an extremely good material for detailed reproduction. Even slight imperfections left after hand finishing the LOM model are faithfully reproduced in the block mould and pieces cast from these moulds.

Whichever AM technology is adopted, such a system saves time in designing and developing tableware, particularly in building a physical

Figure 7.21: Block mould cast from the LOM model of the dinner plate.
Courtesy of Oriental Ceramics Sdn. Bhd., Malaysia

prototype. It can also improve designs by simply amending the CAD model, and the overall system is easy and friendly to use.

As compared to the printing of master patterns for moulding, the direct 3D printing of ceramic or porcelain tableware (plates, bowls, and so on) is still difficult due to the nature of the material. Conventionally, tableware are made from clay before being fired and/or glazed at very high temperatures. Because of this, 3D "ceramic" pottery-making printers used at home usually print only clay, and the user has to manually fire the resulting clay model. There exist at least two commercial 3D printers that print only clay — the LUTUM[44] and the 3D PotterBot,[45] which have been used for vases as well as porcelain tableware.

7.11 Geographic Information System Applications

AM has been applied to create physical models of 3D geographic information system (GIS) objects to replace 2D representations of geographical information. Fig. 7.22 shows a three-dimensional physical model of a city.[46] To be able to do this, CONTEX Scanning Technology introduced its PUMA HS 36 colour scanner. It features iJET Technology and the 18 flatbed colour scanner, which is designed to scan all types of originals, including rare and valuable documents up to A2/C size.

Figure 7.22: A physical model of a city.
Courtesy of CONTEX Scanning Technology A/S

7.11.1 *3D physical map*

The first experiences concerning the reception of the models were very encouraging and surprising; nearly everybody tried to touch the models with his hands, to get a feel in the literal sense.[46] The sensual capability creates a possibility to produce maps for blind and visually impaired persons. The haptic experience could be utilised as an additional stimulus for transmitting a cartographic message and to induce insight.

7.11.2 *3D representation of land prices*

A three-dimensional map visualising the average cost of land in a country or city could be plotted. The height of the prisms would be proportional to the average price of the land. In contrast to a 2D choropleth map depicting value classes, the absolute differences in height could be perceived immediately.[46]

7.11.3 *3D representation of population data*

The volume of the prisms could be proportional to the number of inhabitants in each unit area. The height of the prisms would then be proportional to the population density (see Fig. 7.23).

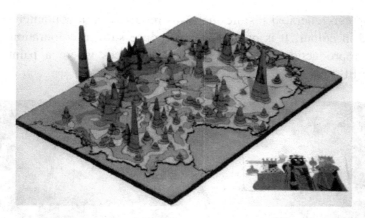

Figure 7.23: Smooth surface depicting population density.
Courtesy of Federal Office for Building and Regional Planning (BBR)

7.12 Art

Although AM technology has been widely used to generate replicas of natural and man-made structures, Neri Oxman, an architect and designer named by Fast Company as one of the "100 Most Creative People" in 2009, is discovering new design and engineering principles through 3D printing to produce complex structures impossible by other fabrication techniques.[47] With 3D printers' capabilities of printing multi-materials, Oxman believes the technology will enable new forms, design freedom and possibilities. Her work has been featured in museums and one of her best known works, Beast, shown in Fig. 7.24, is a sensually curvy chair that can adjust its shape, flexibility and softness to fit each person who sits in it.

Today, AM technology is being used by an increasing number of artists to build a wide variety of sculptural objects. Some of these works are visually realistic and representational, whereas others are abstract. Some of the works created with AM may not have been possible to be made any other way.

Although some abstract objects are the result of pure imagination and artistic free will, they can also be derived solely from mathematics or computation. For instance, as shown in Fig. 7.25, a trefoil torus knot with a computer-generated texture map was rendered on a computer and 3D printed in colour. It is possible to reproduce surface colouration which defies reproduction by hand, such as in spaces where a paint brush cannot fit.

Figure 7.24: Beast, a chair designed by Neri Oxman using 3D printing that responds to the individual's body weight.
Courtesy of Neri Oxman

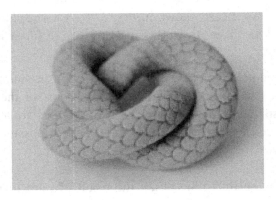

Figure 7.25: Torus knot.
Courtesy of Stewart Dickson

7.13 Architecture

Architectural models are used for visualising the elements of a building or structure considered most important to decision makers. Architectural model making has been revolutionised by AM technology, which has been considered to have replaced traditional techniques because of the following advantages: (a) the ability to create scaled models directly from 3D CAD data in a few short hours. It is currently used by large architectural firms for the creation of proposed and as-built designs. Design changes in particular segments of the construction can be readily rebuilt based on the reproducibility of the AM models and (b) the ability to create multiple models at reasonable cost. For larger projects involving more than one key main contractor, it is advantageous to build multiple copies of the same model for coordination and communication. Fig. 7.26 show an example of architectural models built by 3D printing.

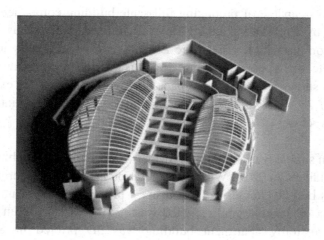

Figure 7.26: An architectural model built by 3D printing.
Courtesy of CONTEX Scanning Technology A/S

7.14 Building and Construction

AM was typically used extensively in building architectural models, but recently the focus has shifted to the delivery of full-scale architectural components as well as elements of buildings such as walls and facades. In construction, every part and structure is unique in size, which means traditional *reductive manufacturing* would require standard sized materials to be removed to fit; *formative manufacturing* would require a modified mould to shape each component.[48] It is hoped that AM processes will bring advantages over conventional manufacturing processes such as (a) the cost-based opportunity to save time and materials by eliminating waste and the need for formwork, (b) the computational design ability providing the freedom to design around individuals and the environment, making each part unique, (c) form optimisation that significantly reduces the quantity of materials needed and (d) AM possessing the potential of building additional functionality into structures.[49]

In 2012, a group of researchers from Loughborough University led the project known as 3D Concrete Printing and Freeform Construction, an AM process capable of producing full-scale building components.[50] The team led by Professor Richard Buswell utilised concrete as its main construction material for the entire AM process, where layers and layers of concrete are deposited in great precision under computer control. Fig. 7.27 shows a 2 m^2 curvy panel being featured in an exhibition, and Fig. 7.28 shows a 1-tonne reinforced concrete bench printed using the AM process.

Other than simple wall structures, AM has been shown to be able to build entire buildings. In 2014, WinSun Decoration Design Engineering Co. of China has built 10 homes in 24 h using a proprietary 3D printer, which uses a mix of concrete and recycled waste.[51] In 2015, WinSun declared that a five-storey apartment building was 3D printed in parts and assembled together with steel reinforcements and insulation.[52]

Singapore's NTU has been at the forefront of the research arena for additively manufactured concrete. NTU has two 3D concrete printers based on different working principles, both of which are approximately 1 × 1 × 1 m in size and can print concrete at a speed of 70–120 mm/s. NTU's researchers are interested in the mechanical characteristics and

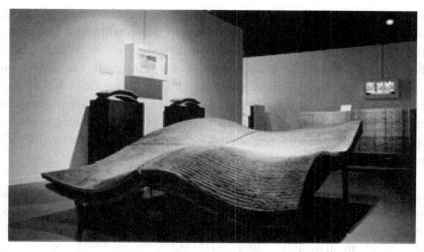

Figure 7.27: A double curved 3D printed concrete panel.
Courtesy of Agnese Sanvito

Figure 7.28: The Wonder Bench 3D printed with concrete.
Courtesy of Loughborough University

performance of the concrete used. Since no supporting formwork is used, the concrete must either have low if not zero slump while still remaining pumpable, or be of a low viscosity while requiring the addition of a chemical accelerator for quick setting once printed. It has also been established that particle shape and granulometric properties have a large impact on compaction behaviour and green strength. Further research in this area has also been done on other types of building material such as environmentally friendly engineered cementitious composites (ECC)[53] and strain-hardened cement-based composites (SHCC).[54]

Additional research has been conducted in areas such as local optimisation of material composition for use in a functionally graded, variable material construction. Craveiro *et al.* has used computer code created by an algorithm editor combined with an FEM software to locally adjust a multi-material ratio according to a Von Mises stress map, in order to reduce the weight and material usage.[55] This is heavily reliant on computer algorithms as well as the physical ability as provided by AM to create walls of multiple materials. This is slightly different from topology optimisation, which changes the exterior shape of the object in an effort to reduce weight or material used. A model is shown in Fig. 7.29, illustrating the variable materials used.

7.15 Fashion and Textile

The fashion industry has started to embrace 3D printing for the past few years. AM's ability to produce complex, traditionally impossible to manufacture forms, coupled with the ease in achieving customisation, is transforming fashion both on the high street and on the runway.

Fashion designers Gabi Asfour, Angela Donhauser and Adi Gil who run the fashion label threeASFOUR have been using 3D printing in their collections known for their biomorphic style. Their newest fall 2016 ready-to-wear collection, called Biomimicry, features garments that take 200 h each to print. AM technology has allowed the trio to create complex and intricately designed geometric patterns drawn from plant geometry and animal anatomy.[56]

Figure 7.29: A model of a conceptual wall prototype of varying spatial porosity.
Courtesy of Flávio Craveiro and Paulo Bártolo

In contrast to the rigid geometric shapes of traditional laser cutting technology, 3D printing gives a softer and delicate cocoon effect. Figs. 7.30 and 7.31 both show 3D printed dresses and shoes designed by threeASFOUR for their spring 2014 collection. The printing of these dresses was outsourced to Materialise, a company reputed for 3D printing haute couture for the catwalk.

Singapore's NTU's School of Mechanical and Aerospace Engineering (MAE) held its first 3D printing festival on 17 December 2013, showcasing fashion pieces made by 3D printers rather than the traditional needle and thread. The international 3D printing competition attracted over 30 entries from 7 countries, and participants created unique and complex designs made possible with AM. The winning entry for the fashion category came from Australia's XYZ designers, who created a fashion piece called the "Hydro-shift top". The network of open spheres

Figure 7.30: 3D printed dress and shoes on display.
Courtesy of threeASFOUR

Figure 7.31: A model displaying a 3D printed dress collection.
Courtesy of threeASFOUR

created a lace-like fabric, which showed a distinct contrast between solid and transparent pieces. The silhouette of the design echoes a traditional Chinese cheongsam, creating a masterpiece combining both tradition and technology. Fig. 7.32 shows the winning 3D printed design by XYZ designers.

The commendation prize went to a group of NTU students comprising three MAE students and four Art, Design and Media students. Their

Figure 7.32: 3D printed winning design "Hydro-shift top".
Courtesy of Nanyang Technological University (NTU) Singapore Centre for 3D Printing (SC3DP)

design cleverly combined both art and technology to form a flowing "chainmail" inspired by the Chinese word for water (水). There are also patterns with large drops of water at the shoulders trickling to smaller drops of water as it unfolds from top to bottom (Fig. 7.33).

AM has also allowed sportswear companies such as Nike, adidas and New Balance to rapidly prototype new shoe designs, resulting in reduced labour hours and slashing the amount of time needed to evaluate a new product. With traditional manufacturing techniques it typically takes New Balance 2 years to go from concept to shipping shoes, but by leveraging AM, many rounds of prototype iterations can be fully tested.[57]

Figure 7.33: 3D printed fashion created by 7 NTU students.
Courtesy of NTU SC3DP

New Balance has also started shipping shoes with 3D printed midsoles, and they hope that AM technology will accelerate to a point where bespoke shoes can be created quickly for each customer, and the materials used are lighter than current materials used.

On top of helping companies streamline their manufacturing process, AM is helping individual consumers gain access to a wide variety of consumer-created designs. Shapeways is an online fashion 3D printing marketplace and community, where users can upload their 3D CAD models to sell as part of their own label or shop.[58] Shapeways will instead be in charge of printing. Users can buy jewellery, shoes, eyewear, pendants, and so on from the many shops on Shapeways, printed in various materials like metals and plastics. CEO Peter Weijmarshausen believes that "the cycle of making a product available and really testing it in the market will go from years to days".[59]

Although AM has been the forefront of fashion and accessories, Tamicare is using the latest AM technology to solve the problem of heavy menstruation for women worldwide. Tamicare uses the proprietary Cosyflex technology to manufacture single-use biodegradable underwear.[60] Their production line machinery sprays fibres and polymers to create stretchable and absorbent fabric. In comparison with traditional underwear production, this cutting edge technology requires no cutting; therefore, almost no waste is produced. Cosyflex-created underwear has recently gone into mass production.

Retail, distribution and manufacturing could very well be on the verge of a new era due to these types of revolutionary technologies. Eventually, customers might be able to co-design and print out bespoke clothes from the comfort of their homes.

7.16   Weapons

AM technology, with its increasing popularity and maturity, offers potentially unlimited possibilities to its users. Its applications have extended to an area where there has been much controversy — 3D printed guns.

The world's first 3D printed gun was introduced in May 2013. Its creator, Cody Wilson from Defense Distributed, took 8 months to create the gun, and then successfully demonstrated it and uploaded its blueprint on the internet.[61] The 3D printed gun, called the "Liberator", is shown in Fig. 7.34.

All 16 pieces, except for the firing pin of the Liberator's prototype, were printed in ABSplus thermoplastic using Stratasys' Dimension SST 3D printer (see Figs. 7.35 (a) and (b)). The gun is able to fire standard handgun rounds.[62]

Using FDM technology, plastic filament flows through a tube to the print head, gets heated to a semi-liquid state and is extruded in thin, accurate layers to create precisely shaped solid objects. The CAD files for the gun designs were downloaded over 100,000 times before the U.S. government intervened to get the files removed from public access.

A few months after the plastic Liberator was introduced, another U.S.-based company called Solid Concepts successfully manufactured and fired the world's first 3D printed metal gun.[63] Most parts of the gun, except for the spring, were made from metal using DMLS, and the gun could withstand the firing of at least 50 rounds. The company made the gun as a testament to prove that 3D printing is a viable solution that can be used in modern manufacturing from metal, and was acquired by Stratasys, Ltd. in 2015.

Three years after the introduction of the handguns, an individual who goes by the pseudonym Derwood created a semi-automatic weapon, which he claims 95% of the parts were 3D printed.[64] The parts that were 3D printed were made from PLA plastic, instead of the regular steel semi-automatic weapon. The barrel, hammer, firing pin, bolts and

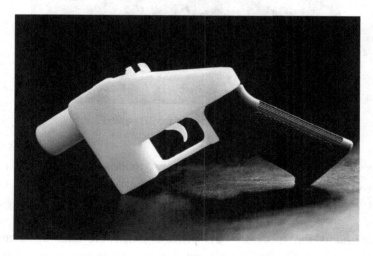

Figure 7.34: The world's first 3D printed gun, the Liberator.
Courtesy of Defense Distributed

Figure 7.35: (a) All 16 pieces except one firing pin printed using ABSplus thermoplastic. Courtesy of Forbes Collection/Michael Thad Carter

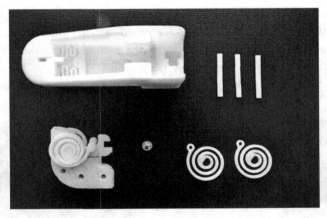

Figure 7.35: (b) Parts of the 3D printed gun.
Courtesy of Defense Distributed

springs were store bought and were made of metal. It is not inconceivable that in the future, fully automatic assault weapons can be 3D printed, in plastic or in metal.

On one hand, the success of 3D printed guns represents an advance in AM technology. On the other hand, the controversies surrounding how easy it is to print an untraceable gun with no serial number from home using only a 3D printer, a computer and Internet connection have shown the dangerous aspect of 3D printing, which until then, had many marvelling at the many opportunities and benefits presented by this technology. Law enforcement agencies worldwide are scrambling to come up with new legislations to curb "undetectable weapons", which are made entirely from plastic and cannot be detected by metal detectors at airports and other sensitive areas. With the significant advances of 3D printing capabilities and accessibility, 3D printed firearms will pose an interesting security threat to public safety.

7.17 Musical Instruments

The art of crafting musical instruments is a skill that takes years to master. A craftsman would design an instrument to create a certain sound tone based on her knowledge and experience using various materials, especially woods of different ages. As the skills and experience of the craftsman improve, so does the quality of her instruments.

The controversial manufacturing of traditional musical instruments using 3D printing has opened up a new horizon for the industry. Traditional designs are now being challenged with new shapes, which integrate the craftsman's (now a computer engineer) complex artwork into the framework of the instruments.

The construction of musical instruments has traditionally been a precise craft, taking many years to master. The strict conformation to traditional standards has also led to little innovation in the design of orchestral instruments. AM, with its ability to manufacture complex forms, can therefore be possibly applied to manufacture intricately shaped instruments, potentially allowing instruments to be mass-produced at a lower cost in the future. In addition, it will allow designers to produce intricate forms not previously possible, leading to more possibilities for tone production.[65]

The technology is already being used to produce novelty electric guitars. Started by Olaf Diegel, a design engineer and professor of mechatronics at Massey University, ODD guitars are a range of electric guitars featuring beautifully intricate, customisable bodies.[66] The Scarab guitar from the range is shown in Fig. 7.36.

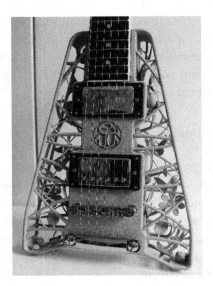

Figure 7.36: The Scarab 3D printed guitar.
Courtesy of Olaf Diegel

Other musical instruments that have been produced using AM include a 3D printed flute and violins. Designed by Amit Zoran, the concert flute in Fig. 7.37 made using Stratasys' PolyJet technology was a proof of concept piece used to study the application of AM to the fabrication of musical instruments.[65]

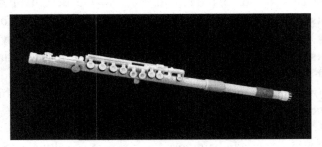

Figure 7.37: A 3D printed flute.
Courtesy of Amit Zoran

EOS produced a replica of a Stradivarius violin body using their technology (see Fig. 7.38). The body is made of EOS PEEK HP3, a high-performance polymer, instead of wood. The violin was then assembled and set up by a luthier with strings, peg box, chin rests, and so on.[67] It is fully functional. The production of a complex instrument, usually made laboriously in a traditional way with traditional materials, via AM is almost blasphemy but it illustrates perfectly the potential of AM technology to change everything.

Figure 7.38: Stradivarius replica by EOS.
Courtesy of EOS GmbH

Eventually, the technology may be used to produce extraordinary instruments that are impossible to fabricate using traditional manu-facturing methods, such as the multi-tube trumpet shown in Fig. 7.39.

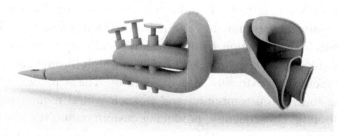

Figure 7.39: Rendering of conceptual multi-tube trumpet.
Courtesy of Amit Zoran

7.18 Food

Applications of AM to designing food constructs was first reported in 2007 by researchers at Cornell University who created an FDM machine called the Fab@Home Model 1.[68] Six systems have been built, of which one is being used in a university engineering course and another exhibited in a museum. The researchers have partnered with the International Culinary Centre in New York City to 3D print chocolate, cheese, cookies and tiny spaceships made of deep-fried scallops, shown in Fig. 7.40. The Model 1 allows the fabrication of multi-material objects with high geometric complexity and sub-millimetre-scale features.

Figure 7.40: 3D printed deep fried scallop in space shuttle form.
Courtesy of Dan Cohen

Another seminal paper by Periard *et al.* claims that the professional culinary community stands to benefit from AM technology as it would enable mass customisation in the industrial culinary sector. Currently, industrial food producers depend heavily on high-throughput processes such as moulding, extrusion and die-cutting. These processes are not amenable to mass customisation and as such would require substantial custom tooling; consequently, producing custom output for low-quantity runs is simply unfeasible. This is where AM's inherent strengths are apparent; AM can produce food with custom, complex geometries while maintaining cost effectiveness. The cost effectiveness is due to the fact that AM does not require custom tooling or extensive manual labour. A potential future application is the custom production of edible giveaways, for example, as marketing collateral for small corporate events.[69]

In 2013, NASA funded the development of a 3D food printer to feed astronauts in space. The agency awarded a Small Business Innovation Research (SBIR) contract and funding of $125,000 to Systems and Materials Research Consultancy, a Texas-based company, to study how to make high-nutrient food using AM technology.[70] The problem with the existing food system is that it fails to meet the nutritional requirements and 5-year shelf life necessary for long-term missions, such as to Mars. It is hoped that a 3D printer can be developed to overcome these challenges, using unflavoured macronutrients like starch and fat, micronutrients, flavours and texture modifiers as the raw materials, ideally without the use of refrigeration because of its high energy consumption. This project will determine if 3D printing can provide a customisable and nutritionally stable variety of foods from shelf stable ingredients while minimising waste and time for space crews. It is also projected that this 3D printer might help the military provide food for soldiers in wartime and on submarines and aircraft carriers.[71]

Looking earthward, a European Union (EU)-funded project by the name of PERFORMANCE hopes to use 3D printing to help patients with dysphagia — chewing and swallowing difficulties often brought about by stroke and dementia.[72] Conventional dysphagia diets are mostly based on puree and pre-mashed food, which is often unappetising and can lead to malnutrition. It is hoped that this project can use industrial 3D printing to create foods that replicate the texture, look and taste of "real" foods, customised to each patient's needs. The food is enriched with specific nutrients based on each patient's size, weight, gender and deficiencies. To help with microwave heating of the food after delivery to nursing homes, PERFORMANCE has developed a split plate with microwave reflectors on top and underneath it for more homogenous heating.

Preliminary results in 2015 showed a generally positive reaction to the 3D printed meals. About 54% of respondents rated the meal's texture as good, 79% found the meals equally heated and 43% would choose the meals in cases of chewing or swallowing difficulties. On top of 3D printing the food, PERFORMANCE wants to collect personalised data from each user about their food preferences, nutritional and vitamin requirements and so on. PERFORMANCE will use this data to 3D print individualised meals on an industrial scale before delivering it to consumers.

3D food printing is drawing the attention of hospital and nursing home administrators such as Gladys Wong, chief dietitian of Khoo Teck Puat Hospital in Singapore, who is embracing 3D printed food as the next step forward in providing safe and nutritious food for patients. She has ordered an in-development 3D food printer from Spain to evaluate the feasibility of implementing this in a hospital environment. Although she finds current puree food satisfactory and tasting "just like regular cooked food", she hopes that 3D printing can bring about a higher level of consistency in the preparation and presentation of meals that are safe and easy to eat for elderly patients and those with chewing and swallowing difficulties. Furthermore, Gladys envisions that 3D printed food will enable personalisation of meals with different levels of nutrients to suit the unique dietary needs of each patient.

7.19 Movies

3D printing has helped the movie industry through the manufacturing of props for use in movies. One of the first examples of AM being used was for *Iron Man 2* (2010), where film production company Legacy Effects created flexible gauntlets no thicker than a dime, allowing them to be worn for a long time without Robert Downey Jr.'s hands getting hot. First, the actor's hands were scanned, and then the metal gloves were 3D printed through DMLS, using an inkjet cartridge to print a layer of powdered plastic, which is then fused with UV light.[73] All these from the designing process to the printing process can be done in a matter of hours. Creating the models by hand would have been too time consuming, but 3D printing allows for an extremely fast way of sculpting and creating concepts.

The main advantage of AM, besides speed, is that one can print out custom-fit costumes, allowing actors to be more comfortable while acting their roles. Downey had previously complained that the original Iron Man suits, which were not 3D printed, were too clunky and uncomfortable to act in.

When filmmakers wanted to blow up an iconic Aston Martin DB5 which appeared in the James Bond film, *Skyfall* (2012), they looked to Propshop Modelmakers, a British company that specialises in the production of film props using a 3D printer by Voxeljet.[74] The car that was blown up was in fact a 3D printed model and the priceless car was spared.

Computer files with the design data for all components were given to Voxeljet to begin the 3D printing process. The models were produced with the layer-by-layer application of particle material that was glued together with a binding agent. A total of 18 individual components for each of the three vehicle models were cleaned and then meticulously assembled and painted to look like the real Aston Martin DB5. This elaborately constructed model was then decimated in flames in the film.

7.20 Marine and Offshore

In marine applications, the use of stainless steel is preferred to other steels such as carbon-manganese and low-alloy steel because of its resistance to weather and environmental corrosion.[75] Stainless steel parts are conventionally manufactured in seven steps: melting and casting, forming, heat treatment, descaling, cutting, finishing and finally end-user post-processing.[76] It is envisaged that 3D printing metals for marine applications will bypass several of these tedious steps while still achieving the required corrosion–resistance properties and mechanical properties such as relative densities, toughness, yield strength, and so on.

The choice of stainless steels within the industry is based on three criteria — resisting stress corrosion cracking, pitting and crevice corrosion, and sour environment.[77] As such, austenitic, super austenitic and duplex (generally containing equal amounts of ferrite and austenite) stainless steels are preferred. SLM can be used for 3D printing these metals and research has been done to find if 3D printed metals have the same properties as casted metals. A review has concluded that relative densities attained by SLM can exceed 99%, as compared to typical densities of 82%–95% for casted components.[77] In general, SLM-produced stainless steels are found to be harder and stronger than conventionally casted stainless steels, but are less ductile and less tough.

Due to the potential benefits accorded by AM, many academic institutions and companies are investing in research in this new technology. The port of Rotterdam will set up its own "AM Fieldlab" with a specific focus on 3D printing metals for the maritime industry, at an investment of several million euros. This follows the success of a 2015 pilot project involving 28 companies and government agencies.[78] Parts tested and printed include hinges, seals, manifolds and propellers.[79]

It is also no surprise that navies around the world are eyeing metal AM technologies for their low cost and fast production times. Naval Sea Systems Command (NAVSEA), the largest of the U.S. Navy's five systems commands, has taken an interest in metal AM for warfighting capabilities. NAVSEA's Naval Surface Warfare Center Dahlgren Division (NSWCDD) and the Combat Direction Systems Activity (CDSA) Dam Neck, Virginia, have jointly demonstrated several metal AM research-and-development proofs of concept, such as building hexapod robots, 3D scanning and printing human heads for custom fit masks and face pieces,[80] and valves for training purposes.[81] Naval Air Systems Command (NAVAIR) has developed a roadmap for integrating AM into some of its products, highlighting the production and flight-testing of 3D printed parts, links and fittings aboard a Bell Boeing V-22 Osprey tiltrotor.[82]

Future challenges in the industry include improving the mechanical properties of 3D printed stainless steels, specifically ductility and toughness, as well as increasing the build size of 3D printed parts and developing new large format AM turnkey systems.

References

1. Mueller, T (1991). Applications of stereolithography in injection molding. *Proc. of the Second International Conference on Rapid Prototyping*, pp. 323–329. Dayton, USA, June 23–26.

2. Wang, JH, CS Lim and JH Yeo (2000). CFD investigations of steady flow in bi-leaflet heart valve. *Critical Reviews in Biomedical Engineering*, 28(1–2), 61–68.

3. Michell, AGM (1904). The limits of economy of material in frame-structures. *Philosophical Magazine*, 8(47), 589–597.

4. Yunus, G, ZW Wong and E Kayacan (2016). Additive manufacturing of unmanned aerial vehicles: current status, recent advances, and future perspectives. *Proc. of the 2nd International Conference on Progress in Additive Manufacturing*, 39–48. doi:10.3850/2424-8967_V02-147

5. Chua, CK, KH Hong and SL Ho (1999). Rapid tooling technology (Part 1) — a comparative study. *International Journal of Advanced Manufacturing Technology*, 15(8), 604–608.

6. Wilkening, C (1996). Fast production of rapid prototypes using direct laser sintering of metals and foundry sand. In *Second National Conference on Rapid Prototyping and Tooling Research*, pp. 153–160. UK, November 18–19.

7. Tsang, HB and G Bennett (1995). Rapid tooling — direct use of SLA moulds for investment casting. In *First National Conference on Rapid Prototyping and Tooling Research*, pp. 237–247. UK, November 6–7.

8. Atkinson, D (1997). *Rapid Prototyping and Tooling: A Practical Guide*. UK: Strategy Publications.

9. Chua, CK, KH Hong and SL Ho (1999). Rapid tooling technology (Part 2) — a case study using arc spray metal tooling. *International Journal of Advanced Manufacturing Technology*, 15(8), 609–614.

10. Venus, AD, SJ Crommert and SO Hagan (1996). The feasibility of silicone rubber as an injection mould tooling process using rapid prototyped pattern. In *Second National Conference on Rapid Prototyping and Tooling Research*, pp. 105–110. UK, November 18–19.

11. Shonan Design Co. Ltd. (1996). *Temp-Less 3.4.3 — UV RTV process for quick development and fast to market.*

12. Schaer, L (1995). Spin casting fully functional metal and plastic parts from stereolithography models. In *The Sixth International Conference on Rapid Prototyping*, 217–236. Dayton, USA, June.

13. Male, JC, NAB Lewis and GR Bennett (1996). The accuracy and surface roughness of wax investment casting patterns from resin and silicone rubber tooling using a stereolithography master. In *Second National Conference on Rapid Prototyping and Tooling Research*, pp. 43–52. UK, November 18–19.

14. Bettay, JS and RC Cobb (1995). A rapid ceramic tooling system for prototyping plastic injection moldings. In *First National Conference on Rapid Prototyping and Tooling Research*, pp. 201–210. UK, November 6–7.

15. Warner, MC (1993). Rapid prototyping methods to manufacture functional metal and plastic parts. *Rapid Prototyping Systems: Fast Track to Product Realization*, 137–144.

16. Lim, CS, L Siaminwe and AJ Clegg (October 1996). Mechanical property enhancement in an investment cast aluminium alloy and metal-matrix composite. In *9th World Conference on Investment Casting*, 17, 1–18. San Francisco, USA.

17. Clegg, AJ (1991). *Precision Casting Processes*. Oxford: Pergamon Press Plc.

18. Venus AD and SJ Crommert (18–19 November 1996). Direct SLS nylon injection. In *Second National Conference on Rapid Prototyping and Tooling Research*, pp. 111–118. UK.

19. Soar, RC, A Arthur and PM Dickens (18–19 November 1996). Processing and application of rapid prototyped laminate production tooling. In *Second National Conference on Rapid Prototyping and Tooling Research*, pp. 65–76. UK.

20. SLM Solutions GmbH (2016). *SLM Metal Powder*. https://slm-solutions.com/ products/accessories-and-consumables/slm-metal-powder

21. Rapid Prototyping and Tooling State of the Industry (2000). *Wohlers Report 2000*. USA: Wohlers Associates Inc.

22. Eyerer, P (1996). Rapid tooling — manufacturing of technical prototypes and small series. *Mechanical Engineering*, 45–47.

23. Stratasys (2015). *Two-person Team Designs 3D-printed Aircraft Through Discovery Method*. http://www.stratasys.com/resources/case-studies/aerospace/selecttech-geospatial

24. Stratasys (2015). *Shaping the Future of Flight*. http://usglobalimages.stratasys.com/ Case%20Studies/Aerospace/CS_FDM_AE_Aurora.pdf?v=635824107576322646

25. Stratasys (2013). *A Turn for the Better*. http://www.stratasys.com/resources/ case-studies/aerospace/kelly-manufacturing/

26. Kellner, T (23 June 2014). *Fit to Print: New Plant Will Assemble World's First Passenger Jet Engine with 3D Printed Fuel Nozzles, Next-Gen Materials*. General Electric Company. http://www.gereports.com/post/80701924024/fit-to-print/

27. Kellner, T (15 July 2014). *World's First Plant to Print Jet Engine Nozzles in Mass Production*. General Electric Company. http://www.gereports.com/post/91763815095/ worlds-first-plant-to-print-jet-engine-nozzles-in/

28. Newman, J (August 2012). *NASA Turns to 3D Printing for Space Exploration Vehicle*. Rapid Ready Technology. http://www.rapidreadytech.com/2012/08/ nasa-turns-to-3d-printing-for-space-exploration-vehicle/

29. NASA (20 April 2016). *3D Printing In Zero-G Technology Demonstration (3D Printing In Zero-G) — 04.20.16*. http://www.nasa.gov/mission_pages/station/ research/experiments/1115.html

30. Wohlers, T (2016). *Wohlers Report 2016: 3D Printing and Additive Manufacturing State of the Industry*. Wohlers Associates, Inc.

31. UTC Aerospace Systems (25 February 2016). *UTC Aerospace Systems Opens Advanced Materials Laboratory in Connecticut*. http://news.utcaerospacesystems.com/2016-02-25-UTC-Aerospace-Systems-opens-advanced-materials-laboratory-in-Connecticut

32. Stratasys (2013). *From Tires to Interiors, Bentley Designs with 3D Printing.* http://www.stratasys.com/resources/case-studies/automotive/bentley-motors

33. Stratasys (2013). *Manufacturing Jigs and Fixtures with FDM.* http://www.stratasys.com/ resources/case-studies/automotive/bmw

34. Available from http://3d.ford.com/

35. Pyper, J (2014). World's First Three-Dimensional Printed Car Made in Chicago. *Scientific American.* http://www.scientificamerican.com/article/world-s-first-three-dimensional-printed-car-made-in-chicago/

36. Available from https://localmotors.com/3d-printed-car/

37. Beck, J (2010). Urbee is the First Car Made by a 3D Printer. *Popular Science.* http://www.popsci.com/cars/article/2010-11/hybrid-car-created-completely-3d-printing/

38. Lee, HB, MSH Ko, RKL Gay, KF Leong and CK Chua (1992). Using computer-based tools and technology to improve jewellery design and manufacturing. *International Journal of Computer Applications in Technology*, 5(1), 72–88.

39. Leong, KF, CK Chua and HB Lee (1994). Finishing techniques for jewellery models built using the stereolithography apparatus. *Journal of the Institution of Engineers, Singapore*, 34(4), 54–59.

40. Solidscape (2014). *Jenny Wu, Revolutionizing Wearable 3D Art.* http://www.solidscape.com/experiences/jenny-wu-revolutionizing-wearable-3d-art/

41. Chua, CK, RKL Gay, SKF Cheong, LL Chong and HB Lee (1995). Coin manufacturing using CAD/CAM, CNC and rapid prototyping technologies. *International Journal of Computer Applications in Technology*, 8(5/6), 344–354.

42. Chua, CK, W Hoheisel, G Keller and E Werling (1993). Adapting decorative patterns for ceramic tableware. *Computing and Control Engineering Journal*, 4(5), 209–217.

43. Chua, CK, RKL Gay and W Hoheisel (1994). A method of generating motifs aligned along a circular arc. *Computers & Graphics: An International Journal of Systems and Applications in Computer Graphics*, 18(3), 353–362.

44. VormVrij LUTUMs, http://www.vormvrij.nl/lutum.php

45. 3D PotterBots. http://www.deltabots.com/printer-selection-guide/

46. Rase, W-D (2002). Physical models of GIS objects by rapid prototyping. In *Symposium on Geospatial Theory, Processing and Applications*. Ottawa.

47. Hill, DJ (2012). *3D Printing is the Future of Manufacturing and Neri Oxman Shows How Beautiful It Can Be.* SingularityHUB. http://singularityhub.com/2012/06/04/3d-printing-is-the-future-of-manufacturing-and-neri-oxman-shows-how-beautiful-it-can-be/

48. Buswell, RA, RC Soar, AGF Gibb and T Thorpe (2007). Freeform construction: mega-scale rapid manufacturing for construction. *Automation in Construction*, 16(2), 224–231. https://dspace.lboro.ac.uk/dspace-jspui/handle/2134/9925

49. Lim, S, *et al.* (2011). Development of a viable concrete printing process. In *The 28th International Symposium on Automation and Robotics in Construction (ISARC2011)*. http://www.iaarc.org/publications/fulltext/S20-3.pdf

50. Lim, S, RA Buswell, TT Le, SA Austin, AGF Gibb, and T Thorpe (2012). Developments in construction-scale additive manufacturing processes. *Automation in Construction*, 21(1), 262–268. https://dspace.lboro.ac.uk/2134/9176

51. Levy, K (16 April 2014). *A Chinese Company 3D Printed 10 Houses in a Day.* Business Insider. Retrieved from http://www.businessinsider.sg/a-chinese-company-3d-printed-10-houses-in-a-day-2014-4/

52. Starr, M (19 January 2015). *World's First 3D-Printed Apartment Building Constructed in China.* CNET. https://www.cnet.com/news/worlds-first-3d-printed-apartment-building-constructed-in-china/

53. Weng, Y, B Lu, MJ Tan and S Qian (2016). Rheology and printability of engineering cementitious composites — a literature review. *Proc. of the 2nd International Conference on Progress in Additive Manufacturing*, pp. 427–432. doi:10.3850/2424-8967_V02-3261

54. Visser, CR and GPAG van Zijl (2007). Mechanical characteristics of extruded SHCC. *Proc. International RILEM Conference on High Performance Fibre Reinforced Cement Composites*, pp. 165–173.

55. Craveiro, F, H Bártolo, JP Duarte, and PJ Bártolo (2016). A strategy to locally optimize the material composition of AM construction elements. *Proc. of the 2nd International Conference on Progress in Additive Manufacturing*, pp. 188–193. doi:10.3850/2424-8967_V02-122

56. Carter, L (16 February 2016). Fall 2016 Ready-to-Wear Threeasfour. *Vogue.* Retrieved from http://www.vogue.com/fashion-shows/fall-2016-ready-to-wear/threeasfour

57. Zaleski, A (15 December 2015). Who's winning the 3D-printed shoe race?. *Fortune.* http://www. http://fortune.com/2015/12/15/3d-printed-shoe-race/

58. Available from http://www.shapeways.com

59. Wolfe, J (19 May 2016). How 3D printing is transforming manufacturing: an interview with shapeways CEO Peter Weijmarshausen. *Forbes.* http://www.forbes.com/sites/joshwolfe/2016/05/19/how-3d-printing-is-transforming-manufacturing-an-interview-with-shapeways-ceo-peter-weijmarshausen

60. Scott, C (23 December 2015). Cosyflex: Tamicare's 3D textile printing technology goes into mass production. 3DPrint.com. https://3dprint.com/112264/tamicare-cosyflex-production/

61. Hadhazy, A (May 2013). Why you should, and shouldn't, worry about the 3D-printed gun. *Popular Mechanics.* http://www.popularmechanics.com/technology/military/weapons/why-you-should-and-shouldnt-worry-about-the-3D-printed-gun-15450141

62. Greenberg, A (May 2013). This is the world's first 3D-printed gun. *Forbes.* http://www.forbes.com/sites/andygreenberg/2013/05/03/this-is-the-worlds-first-entirely-3d-printed-gun-photos/

63. Feinberg, A (11 July 2013). The world just got its first entirely 3D-printed metal gun — and it works. *Gizmodo.* http://gizmodo.com/the-world-just-got-its-first-entirely-3d-printed-metal-1460338036

64. Greenberg, A (3 February 2016). *Someone (Mostly) 3-D Printed A Working Semi-Automatic Gun.* WIRED.com. Retrieved from https://www.wired.com/2016/02/someone-mostly-3-d-printed-a-working-semi-automatic-gun/

65. Zoran, A (2011). The 3D printed flute: digital fabrication and design of musical instruments. *Journal of New Music Research*, 40, 379–387.

66. Odd Guitars (2011). *About.* Retrieved from http://www.odd.org.nz/about.html

67. Lanxon, N (2011). *Hands-on With the EOS 3D-Printed Stradivarius Violin.* http://www.wired.co.uk/news/archive/2011-09/20/3d-printed-stradivarius-violin-eos

68. Malone, E and H Lipson (2007). Fab@Home: the personal desktop fabricator kit. *Rapid Prototyping Journal*, 13(4), 245–255. doi:10.1108/13552540710776197

69. Periard, D, N Schaal, M Schaal, E Malone and H Lipson (August 2007). Printing food. In *Proc. of the 18th Solid Freeform Fabrication Symposium*, pp. 564–574. Austin TX, .

70. 3D Printing: Food in Space (23 May 2013). *NASA.* https://www.nasa.gov/directorates/spacetech/home/feature_3d_food.html

71. NASA SBIR. Proposal 12-1 **H12.04-9357**, Form B-Proposal Summary. http://sbir.gsfc.nasa.gov/SBIR/abstracts/12/sbir/phase1/SBIR-12-1-H12.04-9357.html?solicitationId=SBIR_12_P1

72. Community Research and Development Information Service, European Commission (23 October 2015). 3D-printed food to help patients with dysphagia. http://cordis.europa.eu/news/rcn/124181_en.html

73. Kluang, C (2010). Iron Man 2's Secret Sauce: 3-D Printing. *Fast Company.* http://www.fastcompany.com/1640497/iron-man-2s-secret-sauce-3-d-printing

74. Voxeljet Builds Aston Martin models for James Bond film Skyfall (2012). 3Ders.org.   http://www.3ders.org/articles/20121107-voxeljet-builds-aston-martin-models-for-james-bond-film-skyfall.html

75. Wu, WJ, KF Leong, SB Tor, CK Chua and AA Merchant (2016). State of the art review on selective laser melting of stainless steel for future applications in the marine industry. *Proc. of the 2nd International Conference on Progress in Additive Manufacturing*, pp. 475–481. doi:10.3850/2424-8967_V02-N097

76. Millberg, LS (1992). *Stainless Steel.* http://www.madehow.com/Volume-1/Stainless-Steel.html

77. Geary, EA (2011). *A Review of Performance Limits of Stainless Steels for the Offshore Industry.* Harpur Hill, Buxton Derbyshire.

78. RDM Rotterdam to acquire a "Fieldlab" with 3D metal printers (11 February 2016). *Port of Rotterdam Authority.* https://www.portofrotterdam.com/en/news-and-press-releases/rdm-rotterdam-to-acquire-a-fieldlab-with-3d-metal-printers

79. Pilot Project: 3D printing marine spares final report (25 January 2016). *Port of Rotterdam Authority.* https://www.portofrotterdam.com/en/file/7834/download?token=sexODksY

80. NSWCDD on the "ground floor" of 3D Printing (14 January 2016). *NSWCDD Corporate Communications Division.* http://www.navsea.navy.mil/Media/News/tabid/11975/Article/643045/nswcdd-on-the-ground-floor-of-3d-printing.aspx

81. Navy Officials: 3D Printing to Impact Future Fleet with "On Demand" Manufacturing Capability (23 May 2016). http://www.navsea.navy.mil/Media/News/tabid/11975/Article/778467/navy-officials-3d-printing-to-impact-future-fleet-with-on-demand-manufacturing.aspx

82. Atkinson, P (17 May 2016). NAVAIR Officials Outline Additive Manufacturing Roadmap, Upcoming Parts Test. *Seapower Magazine.*
http://www.seapowermagazine.org/stories/20160517-digital-thread.html

Problems

1. How is the application of additively manufacturing (AM) models related to the purpose of prototyping? How does it also relate to the materials used for prototyping?

2. List the types of industries that AM can be used in. List specific industrial applications.

3. What are the finishing processes that are used for AM models and why are they necessary?

4. What are the typical AM applications in design? Briefly describe each of these applications and illustrate them with examples.

5. What are the typical AM applications in engineering and analysis? Briefly describe each of them and illustrate them with examples.

6. How would you differentiate between the following types of rapid tooling processes: (a) direct soft tooling, (b) indirect soft tooling, (c) direct hard tooling and (d) indirect hard tooling?

7. Explain how an AM pattern can be used for vacuum casting with silicone moulding. Use appropriate examples to illustrate your answer.

8. What are the ways an AM pattern can be used to create injection moulds for plastic parts? Briefly describe the processes.

9. Compare and contrast the use of AM patterns for the following:

    (i)   casting of die inserts,

    (ii) sand casting, and

    (iii) investment casting.

10. What are the AM systems that are suitable for sand casting? Briefly explain why and how they are suitable for sand casting.

11. Compare the relative merits of using LOM parts and SLA parts for investment casting.

12. Explain whether AM technology is more suitable for "high tech" industries like aerospace than it is for consumer product industries like electronic appliances. Give examples to substantiate your answer.

13. Explain how AM systems can be applied to traditional industries like jewellery, coin and tableware.

14. Briefly describe how AM systems can be applied to GIS.

15. What are three advantages of employing AM technology in the aerospace industry?

16. Briefly describe how AM systems can be applied to fashion and textile.

17. Explain the advantages of using AM systems in the movie industry.

18. Explain how multi-material AM can help to reduce weight and material usage in building walls.

19. Explain how 3D food printing can help space agencies achieve long-term human spaceflight. List some of the requirements of such a 3D food printing system.

20. Explain why AM is more suitable for producing parts for unmanned aerial vehicles (UAVs) than conventional manufacturing techniques.

CHAPTER 8

MEDICAL AND BIOENGINEERING
APPLICATIONS

8.1 Planning and Simulation of Complex Surgery

When facing complex operations, such as craniofacial and maxillofacial surgeries, surgeons usually have difficulty figuring out visually the exact location of a tumour or the precise profile of a defect. The precision and speed of a surgical operation depend significantly on the surgeon's prior knowledge of the case and experience. Additive Manufacturing (AM) enables surgeons to practise on a precise model and master the essential details before the actual operation. The hands-on experience helps surgeons achieve better success rates. Hence AM models are frequently used and are vital in surgical planning. The following sections describe some of these applications.

8.1.1 *Congenital malformation of facial bones*

Restoration of facial anatomy is important in cases of congenital abnormalities, trauma or post-cancer reconstruction. In one case, the patient had a deformed jaw at birth, and a surgical operation was necessary to cut out the shorter side of the jaw and alter its position.[1] The difficult part of the operation was the evasion of the nerve canal that ran inside the jawbone. Such an operation was impossible in the conventional procedure because there was no way of visualising the inner nerve canal.

Using a computer-aided design (CAD) model reconstructed from the computerised tomography (CT) images, the position of the canal was identified and a simulation of the amputating process was carried out to determine the actual line of the cut. Furthermore, the use of a semi-transparent resin prototype of the jawbone allowed the visualisation of the internal nerve canal and facilitated the determination of the amputation line prior to the surgery. The end result was a more efficient surgery and vastly improved post-surgery results.

In another case study, a laser digitiser was used instead of CT[2] to capture the external surface profile of a patient with a harelip problem. The triangulated surface of the patient's face was reconstructed in CAD, as seen in Fig. 8.1. Fig. 8.2 shows the stereolithography apparatus (SLA) prototype derived from the CAD data. In this case study, the prototype model provided the validation for the laser scan measurements. In addition, it facilitated an accurate predictive surgical outcome and post-operative assessment of changes in the facial surgery.

8.1.2 *Cosmetic facial reconstruction*

Due to a traffic accident, a patient had a serious bone fracture of the upper and lateral orbital rim in the skull.[3] In the first reconstructive surgery, the damaged part of the skull was transplanted with the shoulder bone. However, shortly after the surgery, the transplanted bone dissolved. Another surgery was required to re-transplant another artificial bone that would not dissolve this time. The conventional procedure of such a surgery would be for surgeons to manually carve the transplanted bone during the operation virtually by "trial and error" until it fitted properly. This operation would have required a lot of time due to the difficulty in carving the bone, which has to be done during the surgery. Using AM, a SLA prototype of the patient's skull was made and then used to prepare an artificial bone that would fit the hole caused by the dissolution. This preparation not only greatly reduced the time required for the surgery, but also improved its accuracy.

In another case at Keio University Hospital of Japan,[4] a 5-month-old baby had a symptom of scaphocephary in the skull. This is a condition that can lead to serious brain damage because it would only permit the

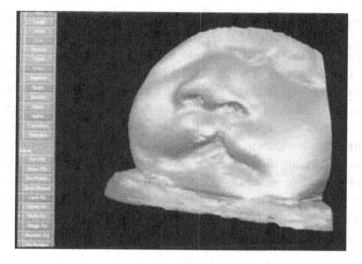

Figure 8.1: CAD model from laser scanner data of a patient's facial details.

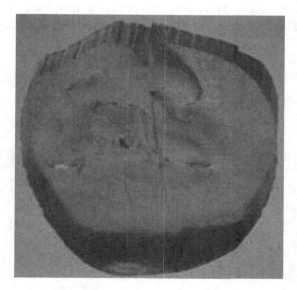

Figure 8.2: SLA model of a patient's facial details.

skull to grow in the front and rear directions. The procedure required was to take the upper half of the skull apart and surgically reconstruct it completely, so that the skull would not suppress the baby's brain as it grows. Careful planning was essential for the success of such a complex operation. This was aided by producing a replica skull prototype with the amputation lines drawn into the model. Next, a surgical rehearsal was carried out on the model with the amputation of the skull prototype according to the drawn lines followed by the reconstruction of the amputated part. In this case, the AM model of the skull provided the surgical procedure with: (1) good three-dimensional (3D) visualisation support for the planning process, (2) an application as training material and (3) a guide for the real surgery.

8.1.3 *Separation of conjoined twins*

On 24 July 2001, two twin sisters were born in a rural village in Guatemala, healthy in every way except for the fact that they were joined at the head[5] (see Fig. 8.3). X-ray pictures showed that the two brains

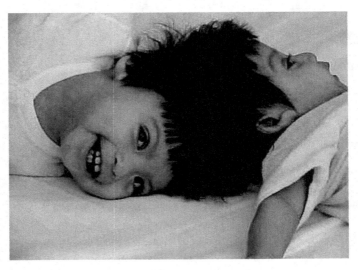

Figure 8.3: Conjoined twins.[5]
Courtesy of Cyon Research Corporation

were separated by a membrane and were otherwise normal in size and had a complete structure. This meant that no brain tissue would have to be cut through during the separation surgery. The arteries that carried oxygenated blood to their brains were also separate, but the veins that drained the blood were interwoven and fed into each other's circulatory systems. The most complex part of the operation was to sort out these veins and reroute each girl's blood supply correspondingly. In this situation, AM had an essential role to play.

A series of CT scans of the two girls were taken. Complicating the task was the fact that the two girls, while connected, could not be arranged in the CT system so that a single scan of their heads could be made. Instead, three sets of scanned data were collected at different angles and then combined into a single 3D model. It took about 3 days to process the CT data and create STL files for AM. Objet's Tempo AM system, made by the Israeli company Objet Geometries Ltd and since merged with Stratasys, was used to construct the skull models of each girl (Fig. 8.4).

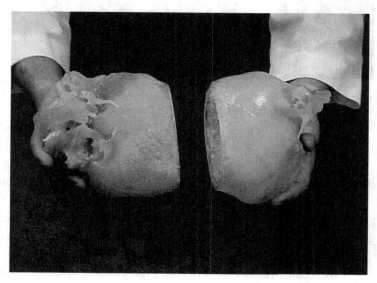

Figure 8.4: Models of each girl's skull.[5]
Courtesy of Cyon Research Corporation

Tempo built the parts by selectively jetting tiny droplets of acrylate photopolymer and then curing the drops, layer-by-layer, with light. It required semi-hardened supports (like a gel) which could be removed simply by jetting water on it. The model of the intersection of the two skulls helped surgeons plan how they would reroute the necessary blood vessels (Fig. 8.5). The operation took about 22 hours to complete. Similar procedures in the past took as long as 97 hours. This significant reduction in time was undoubtedly attributed to the AM model.

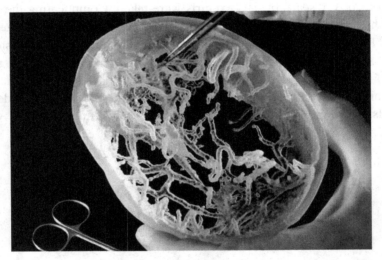

Figure 8.5: A model of the maze of blood vessels where the two skulls connected.[5]
Courtesy of Cyon Research Corporation

8.1.4 *Cancerous brain*

In another case, a patient had a cancerous bone tumour in his temple area, requiring the surgeon to have to access the growth via the front through the right eye socket. The operation was highly dangerous as damage to the brain was likely, which would result in the impairment of some motor functions. In any case, the patient would have lost the function of the right eye.[6] However, before proceeding with the surgery, the surgeon wanted another examination of the tumour location, but this time using a 3D plastic replica of the patient's skull. By studying the model, the surgeon realised that he could reroute his entry through the patient's jawbone, thus avoiding the risk of harming the eye and motor functions. Eventually, the patient lost only one tooth and of course, the tumour. The plastic AM model used by the surgeon was fabricated by SLA from a series of CT scans of the patient's skull.

Other case studies relating to bone tumours have also been reported to have improved success.[7,8] In all cases, not only was the patient spared physical disability as well as the emotional and financial price tags associated with that, but the surgeon also gained valuable insights into his patient using the AM model. From here, non-intuitive alternative improved strategies for the surgery were created by the enhancement to the surgeon's pre-surgical planning stage.

8.2 Customised Implants and Prostheses

For hip replacements and other similar surgeries, these were previously carried out using standardised replacement parts selected from a set range provided by manufacturers based on available anthropomorphic data and the market needs. This works satisfactorily for some types of procedures and patients, but not all. For those patients outside the standard range, in-between sizes, or with special requirements caused by disease or genetics, the surgical procedure may become significantly more complex and expensive. For imperfect fits, these implants may even cause poor gait outcomes and further wear or injury to other joints. AM has made it possible to manufacture a custom prosthesis that precisely fits a patient at a reasonable cost. In relatively short lead time, AM can produce parts with customized geometries and profiles.[9] The following application examples are drawn from a range of applications that span the human anatomy.

8.2.1 *Hip implant*

In one case, a 30-year-old female was diagnosed as having bilateral pseudohypo-chondroplasia with multiple epiphyseal dysplasia.[10] She had a poor range of movement and constant pain in her hips. A total hip arthroplasty with a custom designed femoral implant was recommended. Conventional X-ray (see Fig. 8.6a) showed a very distorted femoral cavities which were wide proximally and extremely narrow in mid-shaft. In order to design a custom stem, CT scans and an AM model (see Fig. 8.6b) were required, which showed a slot type femoral canal. An implant was designed from the CT scan and the model (see Fig. 8.6c). The post-operative X-ray in Fig. 8.6(d) shows that the implant produced a line-to-line fit with the cavity. A clinical follow-up 1 year after the operation showed that the hips were pain-free and functioning well.

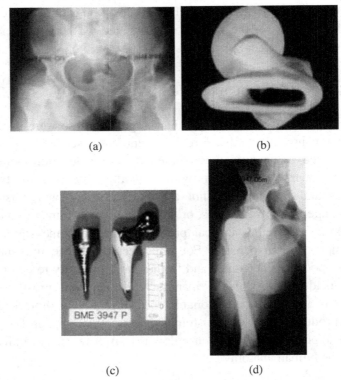

(a)   (b)

(c)   (d)

Figure 8.6: Photographs of total hip arthroplasty. (a) Conventional X-ray showed very distorted femoral cavities which were wide proximally and extremely narrow in mid-shaft. (b) In order to design a custom stem, CT scans and an AM model were required, which showed a slot type femoral canal. (c) An implant was designed from the CT scan and the model. (d) Post-operative X-ray shows that the implant produced a line-to-line fit with the cavity.
Courtesy of Materialise

In another case, a 46-year-old female was diagnosed with malunion of a femoral fracture.[11] She had previous multiple femoral fractures and an osteotomy. Conventional X-ray, as seen in Fig. 8.7(a), showed a misalignment in both A-P and M-L views, and there was an uncertainty on the cross section geometry of the femoral shaft at the level below the

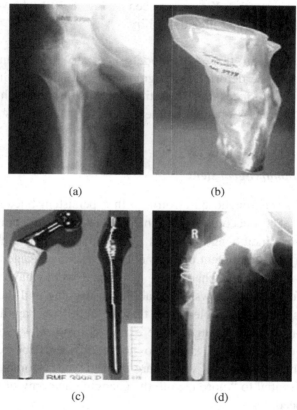

(a) (b)

(c) (d)

Figure 8.7: Total hip arthroplasty for the case of femoral malunion. (a) X-ray showed a misalignment in both A-P and M-L views, and an uncertainty on the cross section geometry of the femoral shaft at the level below the lesser trochanter. (b) The model was made of a transparent resin material so that the internal anatomical geometry of the femoral cavity was clearly visible. (c) A subtrochanteric osteotomy was planned and a custom implant was designed with middle cutting flutes at the region of osteotomy to provide torsional stability. (d) Post-operative X-rays showed that the implant produced a good fit and stability.
Courtesy of Materialise

lesser trochanter. To design a custom implant and determine the form of surgery, CT scans and a medical model were used. The model (Fig. 8.7b) was made of a transparent resin material so that the internal anatomical geometry of the femoral cavity was clearly visible. A subtrochanteric osteotomy was planned and a custom implant (Fig. 8.7c) was designed with middle cutting flutes at the region of osteotomy to provide torsional stability. Post-operative X-rays, as seen in Fig. 8.7(d), showed that the implant produced a good fit and stability.

It has been proven that AM models are very useful in terms of understanding bone deformity, designing implant and planning surgery. Moreover, AM reduces the lead time of custom made hip implant production, decreases operation time and therefore, reduces the risk of infection.[12]

8.2.2 *Buccopharyngeal stent*

A male child was diagnosed at birth with a persistent buccopharyngeal membrane.[13] The buccopharyngeal membrane forms a septum between the primitive's mouth and pharynx. Normally, it completely ruptures during embryo development but was not the case due to a genetic defect. Persistence of the buccopharyngeal membrane would have resulted in the partial fusion of his jaws, the inability in opening and thus speaking as he grows. Another problem was that the child was rapidly growing, and major anatomical changes were expected every 6 weeks up to the age of about 4–5. There was also no readily available commercial buccopharyngeal stent as it was a rare disease, yet customisation of the stent was essential to "morph" with the changing anatomy of growth and surgical procedures.

To solve this problem, the Kandang Kerbau Women's and Children's Hospital in Singapore worked collaboratively with Nanyang Technological University (NTU) Additive Manufacturing Centre (NAMC) and the Rapid Prototyping Laboratory (RPL) at NTU to create a newly designed stent made with biocompatible materials. This material was soft, comfortable, yet rigid enough. The stent was designed to have excellent anti-migration properties when deployed at the pharynx region, yet easy to deploy and extract without causing any trauma to the patient (see Fig. 8.8).

To produce the stent, a master pattern was first fabricated using AM based on the patient's airway morphology. The stent master pattern was then used to create a solid silicone mould. The silicone mould was parted to remove the master pattern, thereby leaving a negative cavity of the implant. The silicone mould was then sprayed with a safety release agent and reassembled. The polyurethane-based resin was then prepared and vacuum cast into the silicone mould under 10^4 Torr in a controlled gas environment. The part was then ejected from the mould and post-processing was carried out. Every 6 weeks, when it was time to produce a new stent, the CAD model for the previous stent was modified using a growth and surgical model semi-empirically developed with the surgeons. The process was repeated until the child reached the age of 5 years old when he was stable enough. With the newly developed stent, the patient was able to breathe, eat, and initiate vocalised sounds as his anatomical structure became more stabilised, thereby learning to talk.

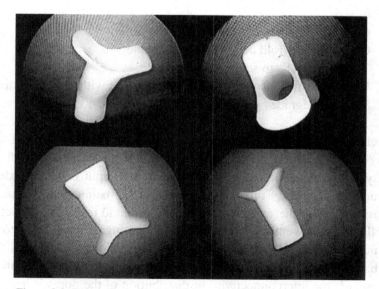

Figure 8.8: Different views of the biocompatible buccopharyngeal stent.

8.2.3 *Tissue engineering scaffolds*

Tissue engineering involves a combination of cells, scaffolds and suitable biochemical factors to improve or replace damaged or malfunctioning organs such as skin, liver, pancreas, heart valve leaflet, ligaments, cartilage and bone. A scaffold is a polymeric porous structure made of biodegradable materials such as polylactic acid (PLA) and polyglycolic acid (PGA). They serve as supports to hold cells.

A typical process in tissue engineering is illustrated schematically in Fig. 8.9 and is as follows:

(1) Examine the defects and determine if a scaffold is necessary.
(2) Isolate functional (undamaged) cells from donor tissue to be cultured.
(3) Select suitable materials to be prototyped.
(4) The patient is introduced to CT or magnetic resonance imaging (MRI) scanning to obtain the geometric data of the defects.
(5) Reconstruct CT data into a virtual model.
(6) Design a scaffold with suitable porous networks to fit the virtual model.
(7) Convert CAD model of scaffold to STL file.
(8) Create the scaffold using an AM machine.
(9) Transplant cells to scaffold.
(10) Implant scaffold to the defected receiver site with growth factors.
(11) Scaffold degrades gradually while cells grow and multiply.

One challenge in tissue engineering is improving the vitality of cells during transplantation. When transplanted to the scaffold, high cell density causes the inner cells to lack nutritional input from the exterior environment, causing some of them to die of malnutrition. Thus the microstructures of the scaffold are very important to the normal functions of cells. The conventional method of fabricating a scaffold is to use organic solvent casting or particulate leaching.[14] However, this has three drawbacks: (1) the thickness is limited; (2) the sizes of the pores are not uniform; (3) the interconnectivity or distribution of the pores is irregular.

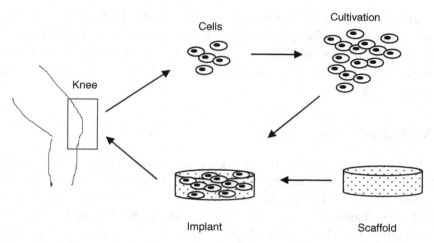

Figure 8.9: An illustration of the tissue engineering process incorporating AM-made scaffolds.

This certainly will affect the number and quality of the cells that are seeding on the scaffold. To solve these problems, AM was introduced.[15-22] Due to its advantage of fabricating intricate objects accurately and quickly, AM has become more and more popular for fabricating tissue engineering scaffolds. Some researchers use AM techniques to study the optimal microstructure of the scaffolds.[23-29] Based on cell type, interconnected networks of the scaffold can be predetermined in a CAD model. After conversion to STL file, a fine structure can be built layer by layer using AM machines. The design of microstructure can be further advanced to make functionally graded scaffolds that closely mimic the structure of a native bone.[30-35] Some researchers apply AM techniques to study scaffold materials, for example, PCL,[36,37] polyHIPE,[38] PEEK/HA composite,[39-42] PVA/HA composite,[40,43,44] PCL/HA composite,[40,45,46] PLLA/HA[47] and PCL/pristine graphene composite,[48] which has been shown to have favourable potential. There are also researchers who work on the indirect fabrication of protein scaffolds such as collagen/gelatine scaffolds[49-53] and silk fibroin scaffolds.[54,55] In this approach, an intricate mould is made by AM first and then used to cast out the protein scaffold. Some researchers are interested in modelling the AM process for a better understanding of scaffold fabrication.[56,57] Other interesting works in this area include the invention of mini-selective laser sintering (SLS) to save materials,[58] AM-made porous scaffold for cardiac tissue engineering[59] and the prevention of bacteria adhesion on scaffolds for bone tissue engineering.[60]

8.2.4 *Bioprinting*

3D Bioprinting, a recently emerging technology, has enabled tissue engineering by depositing living cells and biomaterials to form 3D cells structure.[61] The available technologies for 3D Bioprinting involve extrusion based,[62,63] inkjet based,[62,64] laser based[62,65,66] and integrated technology.[62] These technologies have the potential to engineer body parts and organs using different cells and growth factors, usually accompanied by biomaterials like metals, ceramics, polymers, composites and so on. Hydrogel is one of the popular areas for biomaterial research due to its high water retention property and excellent biocompatibility.[67-71]

On 26 February 2015, the term "bioprinting" has been added into the Oxford Dictionaries along with other new words and phrases, albeit being widely used by scientists in their publications before the recognition. According to Oxford Dictionaries, "bioprinting" is a noun that describes "the use of 3D printing technology with materials that incorporate viable living cells, e.g. to produce tissue for reconstructive surgery".[72]

In contrast to the scaffold method, 3D bioprinting enables automation in tissue engineering methods and promotes the repeatability of cell deposition.[73] Bioprinting also facilitates direct cell to cell interaction as compared to the scaffold approach. Moreover, the bioprinting process can achieve high printing resolution of below 100 microns.[74] However, cells structure printed with AM technology is still inferior in terms of mechanical properties. The bioprinting process is said to be following exactly the process of AM[75]:

(1) Preparation of CAD file
(2) Conversion to STL file format
(3) Slicing of STL file
(4) Tissue printing

One of the many potential applications of bioprinting includes the 4D Bioprinting, in which printed cells or tissues can undergo programmed shape change, size change or/and pattern change with time.[76] On the other hand, a company known as Modern Meadow aims to print leather, and ultimately meat, as an attempt to reduce the need of slaughtering livestock. Meanwhile, a trend of research on *in vivo* bioprinting is observed in which tissues are repaired or fabricated directly on the defective site of living body.[77]

Despite being proposed as "organ printing",[78,79] the current practical usage of bioprinting is limited to simple tissues that lack complex structures.[80] One of the examples of bioprinting's applications is *in vitro* drug testing and discovery.[81,82] Bioprinting allows the screening of drug by using 3D-printed tissues of specific characteristics instead of using human or animals as the drug testing subject. The challenge of tissue vascularization,[83–85] as well as innervation, needs to be overcome before fully functional complex organs, for example, heart, liver and kidney can be achieved.

8.3 Design and Production of Medical Devices

AM has impacted the biomedical field in several important ways. Besides biomodelling for surgical planning and medical implants, another obvious application is to design, develop and manufacture medical devices and instrumentation. AM technology is also being used to fabricate drug dosage forms with precise and complex time release characteristics. In addition, the market value of new designs of medical devices or instruments can be proved with the help of AM.

8.3.1 *Biopsy needle housing*

Biomedical applications are extended beyond design and planning purposes. The prototypes can serve as a master for tooling, such as a urethane mould. At Baxter Healthcare, a disposable-medical-products company, designers rely on two AM processes, SLA and solid ground curing (SGC), to create master models from which they develop metal castings.[6] The master models also serve as a basis for multiple sub-tooling processes. For example, after a master model has been generated via one of the AM machines, the engineers might build a urethane mould around it, cut open the mould, pull out the master, and then inject thermoset material into the mould to make prototype parts.

This process is useful in situations where multiple prototypes are necessary because the engineers can either reuse the rubber moulds or make many moulds using the same master model. The prototypes are then delivered to customer focus groups and medical conferences for professional feedback. Design changes are then incorporated into the master CAD database. Once the design is finalised, the master database is used to drive the machining of the part. Using this method, Baxter Healthcare has made models of biopsy needle housing and many other medical products.

8.3.2 *Drug delivery device*

Drug delivery refers to the delivery of a pharmaceutical compound to humans or animals. The method of delivery can be classified into two ways, invasive and non-invasive. Most of the drugs adopt non-invasive ways by oral administration. Other medications, however, cannot be delivered by oral administration due to its ease of degradation, for example, proteins and peptide drugs. Generally they are delivered by the invasive method of injection.

Currently many research efforts are focusing on target delivery and sustained release formulation. Target delivery refers to delivery drugs at the target site only, for example, cancerous tissues, and not elsewhere along the route. Sustained release formulation refers to releasing the drug over a period in a controlled manner, thus achieving optimally therapeutic concentration.

Polymeric drug delivery devices (DDDs) play an important role in sustained release formulation. However, the current fabrication methods lack precision, which impairs the quality of the device, resulting in a decrease in the efficiency and effectiveness of drug delivery. The SLS process is explored to fabricate a polymeric matrix of drugs layer by layer using powdered materials.[86–90] The capability to build controlled released DDD is also demonstrated in a study conducted by the Nanyang Technological University. In this research, a varying-porosity circular disc with a denser outer region that acts as a diffusion barrier region, and a more porous inner region that acts as a drug encapsulation region is fabricated using the SLS process (see Fig. 8.10). Biodegradable powder materials are used. This research also concludes that control of SLS process parameters, such as laser power, laser scan speed and bed temperature will influence the porous microstructure of the polymeric matrix. Because the SLS process is not capable of processing two or more materials separately, research exploring the possibilities of using SLS to perform a dual material operation is therefore being carried out. In this research, two processes — space creation and secondary powder deposition — are integrated to form a foundation for future work in multi-material polymeric DDD fabrication.[91,92]

Figure 8.10: Polymeric cylindrical DDD built with the SLS.

8.3.3 *Masks for burn victims*

Burn masks are plastic shields worn by patients with severe facial burns. They are used as treatment devices to prevent scar tissue from forming. Burn patients are required to wear the mask for at least 23 hours a day between tissue reconstructive surgery sessions.

The traditional methods for producing burn masks vary and are never simple. The most conventional way is to form a mould by applying a plaster material to the patient's face. The process can be extremely anxiety-provoking for the patient since their eyes and mouth are completely covered throughout the drying process. The weight of the plaster can shift the tissue on the face, making it virtually impossible to get an exact replication.

AM can be used to produce custom fit masks for burn victims. The process begins by digitising the patient using non-contact optical scanning. The scan data are used to produce an AM model of a mask that is a very precise representation of the patient's facial contours. The mask applies pressure to the face to slow the flow of blood to the healing skin. This, in turn, reduces the formation of scar tissue.

A 5-year-old child and his family flew from their Colorado home to have a burn mask developed[93] (see Fig. 8.11). Prior to their visit, physicians

Figure 8.11: A burn mask for a child developed by Total Contact.
Courtesy of Total Contact

and occupational therapists had developed two burn masks using the conventional plaster method for the child, but encountered unsatisfactory results. The AM method enabled the child and his family to return to Colorado in less than 48 hours with a complete, more accurately fitting mask. Today the child's soft tissue is healing better due to the accuracy of the burn mask, which creates smoother skin surfaces and causes less abnormal scarring.

8.3.4 *Functional prototypes help prove design value*

Honeywell is one of the world's leading companies in militarised head-mounted display technology.[94] The display technology has also made minimally invasive surgeries easier. In actual surgeries, the functional prototype headset was worn by military surgeons, replacing the need for conventional cathode ray tube (CRT) monitors.

At the initial stage of developing the new miniature colour-display technology, a prototype was not used. However, the research scientists found that without a prototype they could not get meaningful feedback on their design. On the other hand, they could not justify the tooling costs for making a real product. They then turned to a design firm for the AM of such a model. To allow the surgeons to provide truthful feedback, the prototype had to look and feel like the real product. A functional and durable material was required. After careful consideration, ABS was selected because the continuous copolymer phase gave the materials rigidity, hardness and heat resistance. In addition, it was easy to finish and approximate the injection-moulded plastics well using AM to build. For the AM machine, fused deposition modelling (FDM) was chosen for the electronic covers and case housings because the operating temperatures were to be high. For example, the electronics pack that supplies power to the displays was very small and can generate a great deal of heat in its prototype form. It contained a very high-density printed circuit board with the electronic equivalent of two television sets, plus video converters.

The final result was very pleasing. With the prototype, surgeons saw the benefits immediately and were quite certain that the newly developed display technology had proven its market value. AM has made a difference by being able to sell the concept to medical market leaders. The alternative could have resulted in a loss of funding for the programme due to communication gaps and a lack of understanding about the technology.

8.4 Forensic Science and Anthropology

Surgeons are not the only people interested in bones. Anthropologists and forensic specialists share the same interest, too. AM and related technologies have enabled these scientists to put a recognisable face on skeletons and share precious replicas of rare finds.

8.4.1 *The mummy returns*

Imagining how ancient people look different from the humans of today is made viable to us by AM. Viewing the faces of people some 3000 years ago through their skeletal finds is basically restoring history in real time.

In one study, an ancient Egyptian mummy from the collection of the Egyptian Museum (in Torino, Italy) was selected as a research example. This mummy was well-preserved and completely wrapped.[95] Without having to unravel the mummified wrappings, CT scans were obtained to establish geometric data. CT is the most important non-invasive imaging technique for obtaining fundamental data for 3D reconstructions of the skull and the body, especially with wrapped mummies. Based on the CT scan data, a 3D nylon model was built using the SLS technology, which provided the frame for facial reconstruction of the mummy.

Anthropometric data, the conditions of the remaining dehydrated tissues, and the most accepted scientific and anthropological criteria were used to restore the physical appearance of the mummy. This was done by progressive layering plasticine onto the nylon model.

8.4.2 *The beer bottle*

Early on a Sunday in Kingsbridge, young 17-years-old footballer Alex was in dispute with a drunken father-of-three, Lee Dent. Just after 2 a.m., Lee Dent swung a beer bottle into Alex's throat, leaving a seven and a half centimetres deep wound which cuts through Alex's jugular vein and artery. Alex was pronounced dead due to excessive blood loss.

Hours after Alex's death, Lee Dent gave himself up to the police. However, he denied his act of murder, but instead, claimed that the incident was an act of self-defence against Alex, who seemed to hold a knife. Lee Dent also claimed that the murder was an accident, not knowing that he was holding a beer bottle in his hand when he tried to punch Alex. The theory of self-defence is questionable when prosecutors claimed that Lee Dent fabricated the story by cutting himself to back the account of Alex having a knife. Furthermore, Lee Dent's claim of accident was not believed by prosecutors, thus leaving the request of having Lee Dent to demonstrate the way he held the beer bottle during the scene. However, the demonstration could not be done using the original glass beer bottle as it is ill-advised to hand over a "weapon" to a murder suspect while giving evidence in the court.

In this case, AM technology came to be a great assistance for the murder conviction. Prosecutors requested assistance from the Plymouth City College to 3D print the polymer replica of the murder weapon — a Newcastle Brown Ale glass beer bottle. The polymer beer bottle replica was printed with a CubeX 3D printer, spending 28 hours of time. Thanks to the replica, demonstration of bottle holding in the court was safely conducted, allowing for cross-examination on Lee Dent's claim of self-defence. Lee Dent was found to be guilty later and was sentenced to life in prison with a minimum of 22 years' term.

8.4.3 *The woman in the river*

In another criminal case, Laminated Object Manufacturing (LOM) was used[96] to perform forensic reconstruction of a murdered body. A dismembered body of a woman was uncovered in rural Wisconsin. However, it was impossible for the police to identify who she was because the skin of the face had been removed. Fortunately, the skull of the victim was preserved whole. Using CT scanning, a virtual model of the skull was constructed. After conversion to STL file, a model of the victim was created layer-by-layer using the LOM machine (Fig. 8.12). Forensic anthropologists performed a facial reconstruction directly on the LOM model, and photographs of the reconstructed face were distributed for identification (see Fig. 8.13). One response gave clues to search for the suspect. In the end, the murderer was arrested.

Figure 8.12: (left) LOM model before decubing and (right) finished model.

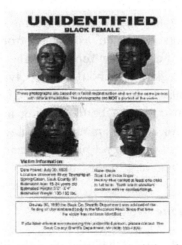

Figure 8.13: The flyer distributed for identification, depicting the completed AM-assisted reconstructed face with a variety of hairstyles.

8.5 Visualisation of Bio-Molecules

8.5.1 *Bio-molecular models for educational purposes*

The stereo-structure of a molecule determines its physical and chemical properties. In biological sciences, the structures of proteins are intensively studied in order to understand biological activities. Researchers generally have a good spatial sense to imagine the structure using computerised virtual construction. However, to clearly explain it to students is very challenging, because words are always vague to some degree and each student has his own perception and understanding.

AM technology can be used to produce accurate, 3D physical models of proteins and other molecular structures. From protein data banks where proteins of known structures are stored, the structure file of the protein of interest can be found. Based on xyz coordinates obtained from the structure file, a CAD model could be constructed. After conversion to STL file, a 3D protein model can be rapidly prototyped (see Fig. 8.14).

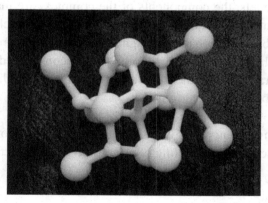

Figure 8.14: AM model of a protein molecule made using Stratasys' PolyJet™ system.

Students are able to interact with the model and explore its shape and chemical properties. Colour schemes can be used to identify the helices and sheets.

8.5.2 *3DP for modelling protein-protein interactions study*

Protein-protein interactions refer to the association and disassociation of protein molecules. They play a vital role for nearly every biological process in a living cell. A good understanding of the mechanisms of these interactions can help clarify the cause and development of disease and provide new therapeutic approaches.

One issue in the study of protein-protein interactions is to deduce the spatial arrangement of the complex (a protein carrying another one for a long time or a brief interaction just for modification); in other words, to predict the molecular structure of the protein complex using the known protein structures.[97] In principle, especially for a large complex, the interface of interaction is unique due to surface complementarity. This is where physical models come in and work.

Structures of the known proteins can be found in Protein Data Bank, thus xyz coordinates can be obtained to build CAD models. After conversion to STL file, accurate physical models are made using AM machines. Then these models, the components of the complex, are manoeuvred by hand using translational and rotational orientations to quickly search for the most suitable and probable configuration. Based on the configuration identified by the models, a virtual complex is reconstructed in molecular graphics programs. After modification, the structure of the complex is considered to be finally determined.

In 2002, an investigative study proved that 3D-Rapid Prototyping is in general eligible for the modelling of protein-protein interaction.[98]

8.6 Bionic Ear

A team of researchers from Princeton University led by Michael McAlpine made history when they successfully harnessed the power of AM to create a "bionic ear".[99] Fig. 8.15 shows the bionic ear manufactured using a 3D printer. This bionic ear is capable of detecting frequencies a million times higher than the normal range of human hearing. The bionic ear is an evidence of how 3D printing can seamlessly integrate two materials, electronic and biological tissues, that traditionally do not blend well together due to the former being rigid and fragile, and the latter being soft and flexible.

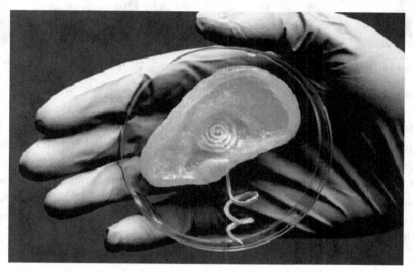

Figure 8.15: The bionic ear printed using a 3D printer has an integral spiral antenna that resembles the ear's cochlear structure.
Courtesy of the American Chemical Society

Using a computer model of an ear and an internal antenna coil connected to an external electrode, the 3D printer alternates among three "inks": a mix of bovine cartilage-forming cells suspended in a thick goo of hydrogel, a suspension of silver nanoparticles to form the coil and external cochlea-shaped electrodes, and silicone to encase the electronics (see Fig. 8.16). The whole printing process takes about 4 hours, where the ear is then cultivated in a nutrient-rich liquid to allow for cell growth producing collagen and other molecules. The original surroundings are then replaced with cartilage.[100]

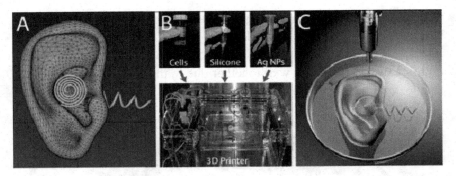

Figure 8.16: The bionic ear computer model and the 3 "inks" required for 3D printing the bionic organ.
Courtesy of the American Chemical Society

The ear is not able to replace the human ear but it was created to demonstrate the power of 3D printing technology in cybernetics, which combines biological tissue with electronics. Further research could lead to synthetic replacements for actual human functions and create additional senses to humans' five traditional senses. 3D printing may also address the main tissue engineering challenges of building organs with blood vessels.

8.7 Dentistry

As the world enters a new era of manufacturing revolution with 3D printing, the dental industry is showing an interest in 3D printing with a general shift towards digital dentistry aided by oral scanners. Using oral scanners combining CAD/CAM and 3D printing, dentists and lab owners are able to customise and speed up the production of crowns, bridges, stone models and a range of orthodontic appliances with quality and precision.

8.7.1 *Speeding up the production of crowns and bridges*

Dental technicians used to prepare every crown and bridge manually, which took a very long period of time. The patient was required to bite into a tray of gooey silicone at the dentist's office to make the models of the patient's teeth. After hardening, the impression was sent to the lab, where gypsum was poured on it to create the model. The entire process could take several weeks. AM is well positioned in dentistry by speeding up the entire process of making crowns and bridges, which are typically high-end products and require customisation to suit different patients' needs.

Highland Dental Laboratory, a full service dental lab based in Texas, USA utilises EnvisonTEC's Perfactory® DDP4 3D printer to produce crowns and bridges. Previously, crowns were pressed or cast manually by hand, which was labour intensive and costly. In addition, a bulk of smaller and less complicated work was outsourced to cope with the demand. With the installation of the 3D printer, the lab was able to produce crowns of e.max, zirconia, full cast alloy and other materials in-house, thus giving the company a vastly significant improvement in quality control and job turnarounds.[101] Instantaneous data transferred from dentists straight to the lab help speed up the production of the model to the finished product.

8.7.2 *3D printed custom-made toothbrush*

The first toothbrush was invented in the 1780s, made from cattle bone and featured a swine bristle brush. Aside from basic modifications, the form of the humble toothbrush has remained unchanged for centuries but now, with AM technology, the way we brush our teeth will change forever. The Blizzident is a 3D printed toothbrush that is custom made from a scan taken by a dentist and tailor-fit for each individual customer.[102] The customised 3D mouthpiece looks like a set of dentures filled with hundreds of bristles (see Fig. 8.17). It requires the customer to wear it in the mouth and uses biting and chewing motions to clean the teeth and gums in a matter of seconds. The Blizzident is not just a time-saving device but it also eliminates human errors of manual brushing. With plenty of soft bristles angled to clean the teeth, gums and even the tongue, the Blizzident will be able to brush all the traditionally difficult-to-reach places, giving the user a thoroughly cleaned mouth.

Figure 8.17: The Blizzident toothbrush is 3D printed to custom fit a customer's mouth and teeth profile.
Courtesy of Blizzident

To make the toothbrush, an impression of the teeth is scanned and uploaded to the company's server. The makers find the optimal placement of 400 bristles by performing simulations of biting and chewing motions and create a CAD model of the brush. The toothbrush is then 3D printed by high precision machines using SLA, which employs a vat of liquid ultraviolet curable photopolymer "resin" and an ultraviolet laser to build the parts layer-by-layer. The materials used to 3D print these devices are made from the same specially classified biocompatible printing materials used to form implantable medical devices. The company has also taken steps to ensure the materials and printers used are able to create totally smooth surfaces for perfect hygiene. Ordinary 3D printers produce rough surfaces where dirt can hide and get trapped easily. The Blizzident is a perfect example of what 3D printing can do in creating customised versions of everyday objects.

8.7.3 *Digital dentistry and orthodontics*

Traditional dentistry relies heavily on devices like stone moulds and braces to restore dental structure and health while digital dentistry relies on innovative technologies such as oral scanners and 3D printing. In 1997, a start-up company approached 3D Systems with a novel idea to leverage digital scanning and 3D printing to dramatically change the way teeth were straightened for orthodontic patients. This company believed that personal orthodontic treatment devices manufactured using the latest 3D printing technology could eliminate the reliance on metal or ceramic brackets and wires, thereby drastically reducing the aesthetic, discomfort and other limitations associated with braces. Align Technology is the company which invented Invisalign Systems. They worked closely with 3D Systems, who helped them develop a customised solution combining digital scanning and 3D printing technology.[103] Stereolithography technology is used to make the tools to manufacture numerous clear plastic aligners, which are worn sequentially by patients to move teeth into a desired final orientation. Being the early adopter of 3D printing for mass customisation has positioned Invisalign System as the market leader of invisible orthodontic system, generating over $500 million in revenue globally in 2012.

References

1. Kaneko, T, M Kobayashi, Y Tsuchiya, T Fujino, M Itoh, M Inomata, M Uesugi, K Kawashima, T Tanijiri and N Hasegawa (1992). Free surface 3-dimensional shape measurement system and its application to Mictotia ear reconstruction. In *The Inaugural Congress of the International Society for Simulation Surgery.*

2. Chua, CK, SM Chou, WS Ng, KY Chow, ST Lee, SC Aung and CS Seah (1998). An integrated experimental approach to link a laser digitiser, a CAD/CAM system and a rapid prototyping system for biomedical applications. *International Journal of Advanced Manufacturing Technology,* 14(2), 110–115.

3. Adachi, J, T Hara, N Kusu and H Chiyokura (1993). Surgical simulation using rapid prototyping. In *Proceedings of the Fourth International Conference on Rapid Prototyping,* pp. 135–142.

4. Koyayashi, M, T Fujino, H Chiyokura and T Kurihara (1992). Preoperative preparation of a hydroxyapatite prosthesis for bone defects using a laser-curable resin model. In *The Inaugural Congress of the International Society for Simulation Surgery.*

5. Cyon Research Corporation (2006). *Rapid Prototyping Helps Separate Conjoined Twins.*
http://www.newslettersonline.com/user/user.fas/s=63/fp=3/tp=47?T=open_article,4 84858&P=article

6. Mahoney, DP (1995). Rapid prototyping in medicine. *Computer Graphics World,* 18(2), 42–48.

7. Swaelens, B and JP Kruth (1993). Medical applications in rapid prototyping techniques. In *Proceedings of the Fourth International Conference on Rapid Prototyping*, pp. 107–120.

8. Jacobs, A. B Hammer, G Niegel, T Lambrecht, H Schiel, M Hunziker and W Steinbrich (1993). First experience in the use of stereolithography in medicine. In *Proceedings of the Fourth International Conference on Rapid Prototyping*, pp. 121–134.

9. Lim, CS, P Eng, SC Lin, CK Chua and YT Lee (2002). Rapid prototyping and tooling of custom-made tracheobronchial stents. *International Journal of Advanced Manufacturing Technology*, 20(1), 44–49.

10. Materialise NV (2007).
http://www.materialise.be/medical-rpmodels/case6_ENG.html

11. Materialise NV. (2007).
http://www.materialise.be/medical-rpmodels/case7_ENG.html

12. Popovich, A, V Sufiiarov, I Polozov, E Borisov and D Masaylo (2016). Producing hip implants of titanium alloys by additive manufacturing. *International Journal of Bioprinting,* 2(2), 78–84. doi: 10.18063/IJB.2016.02.004

13. Tan, SS., HK Tan, CS Lim and WM Chiang (2006). A novel stent for the treatment of persistent buccopharyngeal membrane. *International Journal of Pediatric Otorhinolaryngology.* 70(9), 1645–1649.

14. Yang, S *et al.* (2001). The design of scaffolds for use in tissue engineering. Part I. Traditional factors. *Tissue Engineering*, 7(6), 679–689.

15. Yang, S *et al.* (2002). The design of scaffolds for use in tissue engineering. Part II. Rapid prototyping techniques. *Tissue Engineering*, 8(1), 1–11.

16. Leong, KF, CM Cheah and CK Chua (2003). Solid freeform fabrication of three-dimensional scaffolds for engineering replacement tissues and organs. *Biomaterials*, 24(13), 2363–2378.

17. Yeong, WY *et al.* (2004). Rapid prototyping in tissue engineering: challenges and potential. *Trends in Biotechnology*, 22(12), 643–652.

18. Bártolo, PJ *et al.* (2009). Biomanufacturing for tissue engineering: present and future trends. *Virtual and Physical Prototyping*, 4(4), 203–216.

19. Derby, B (2012). Printing and prototyping of tissues and scaffolds. *Science*, 338(6109), 921–926.

20. Chua, CK, WY Yeong and KF Leong (2005). Rapid prototyping in tissue engineering: A state-of-the-art report. In *Virtual Modelling and Rapid Manufacturing — Advanced Research in Virtual and Rapid Prototyping*, Bártolo (ed.). London Taylor & Francis Group.

21. Chua, CK, MJJ Liu and SM Chou (2012). Additive manufacturing-assisted scaffold-based tissue engineering. In Innovative Developments in Virtual and Physical Prototyping. In *Proceedings of the 5th International Conference on Advanced Research and Rapid Prototyping*.

22. Boland, T *et al.* (2007). Rapid, prototyping of artificial tissues and medical devices. *Advanced Materials and Processes*, 165(4), 51–53.

23. Chua, CK *et al.* (2003). Development of a tissue engineering scaffold structure library for rapid prototyping. Part 1: Investigation and classification. *International Journal of Advanced Manufacturing Technology*, 21(4), 291–301.

24. Chua, CK *et al.* (2003). Development of a tissue engineering scaffold structure library for rapid prototyping. Part 2: Parametric library and assembly program. *International Journal of Advanced Manufacturing Technology*, 21(4), 302–312.

25. Chua, CK *et al.* (2003). Novel method for producing polyhedra scaffolds in tissue engineering. In *Virtual Modelling and Rapid Manufacturing — Advanced Research in Virtual and Rapid Prototyping*.

26. Cheah, CM *et al.* (2004). Automatic algorithm for generating complex polyhedral scaffold structures for tissue engineering. *Tissue Engineering*, 10(3–4), 595–610.

27. Naing, MW *et al.* (2005). Fabrication of customised scaffolds using computer-aided design and rapid prototyping techniques. *Rapid Prototyping Journal*, 11(4), 249–259.

28. Cai, S, J Xi and CK Chua (2012). A novel bone scaffold design approach based on shape function and all-hexahedral mesh refinement. *Methods in Molecular Biology*, 868, 45–55.

29. Ang, KC *et al.* (2006). Investigation of the mechanical properties and porosity relationships in fused deposition modelling-fabricated porous structures. *Rapid Prototyping Journal*, 12(2), 100–105.

30. Chua, CK *et al.* (2010) *Process flow for designing functionally graded tissue engineering scaffolds.*

31. Chua, CK *et al.* (2011). Selective laser sintering of functionally graded tissue scaffolds. *MRS Bulletin*, 36(12), 1006–1014.

32. Sudarmadji, N *et al.* (2011). Investigation of the mechanical properties and porosity relationships in selective laser-sintered polyhedral for functionally graded scaffolds. *Acta Biomaterialia*, 7(2), 530–537.

33. Sudarmadji, N, CK Chua and KF Leong (2012). The development of computer-aided system for tissue scaffolds (CASTS) system for functionally graded tissue-engineering scaffolds. *Methods in molecular biology (Clifton, N.J.)*, 868, 111–123.

34. Chua, CK, N Sudarmadji and KF Leong (2007). Functionally graded scaffolds: The challenges in design and fabrication processes. In *Proceedings of the 3rd International Conference on Advanced Research in Virtual and Rapid Prototyping: Virtual and Rapid Manufacturing Advanced Research Virtual and Rapid Prototyping.*

35. Leong, KF *et al.* (2008). Engineering functionally graded tissue engineering scaffolds. *Journal of the Mechanical Behavior of Biomedical Materials,* 1(2), 140–152.

36. Wang, H, S Vijayavenkataraman, Y Wu, Z Shu, J Sun and JFY Hsi (2016). Investigation of process parameters of electrohydro- dynamic jetting for 3D printed PCL fibrous scaffolds with complex geometries. *International Journal of Bioprinting*, 2(1), 63–71. doi: 10.18063/IJB.2016.01.005

37. Leong, WS, SC Wu, KW Ng and LP Tan (2016). Electrospun 3D multi-scale fibrous scaffold for enhanced human dermal fibroblasts infiltration. *International Journal of Bioprinting*, 2(1), 81–92. doi: 10.18063/IJB.2016.01.002

38. Malayeri, A, C Sherborne, T Paterson, S Mittar, IO Asencio, PV Hatton and F Claeyssens (2016). Osteosarcoma growth on trabecular bone mimicking structures manufactured via laser direct write. *International Journal of Bioprinting*, 2(2), 67–77. doi: 10.18063/IJB.2016.02.005

39. Tan, KH *et al.* (2003). Scaffold development using selective laser sintering of polyetheretherketone-hydroxyapatite biocomposite blends. *Biomaterials*, 24(18), 3115–3123.

40. Tan, KH *et al.* (2005). Selective laser sintering of biocompatible polymers for applications in tissue engineering. *Bio-Medical Materials and Engineering*, 15(1–2), 113–124.

41. Tan, KH *et al.* (2005). Fabrication and characterization of three-dimensional poly(ether-ether-ketone)/-hydroxyapatite biocomposite scaffolds using laser sintering. *Proceedings of the Institution of Mechanical Engineers, Part H: Journal of Engineering in Medicine*, 219(3), 183–194.

42. Vaezi, M and S Yang (2015). A novel bioactive PEEK/HA composite with controlled 3D interconnected HA network. *International Journal of Bioprinting*, 1(1), 66–76. doi: 10.18063/IJB.2015.01.004

43. Chua, CK *et al.* (2004). Development of tissue scaffolds using selective laser sintering of polyvinyl alcohol/hydroxyapatite biocomposite for craniofacial and joint defects. *Journal of Materials Science: Materials in Medicine*, 15(10), 1113–1121.

44. Wiria, FE *et al.* (2008). Improved biocomposite development of poly(vinyl alcohol) and hydroxyapatite for tissue engineering scaffold fabrication using selective laser sintering. *Journal of Materials Science: Materials in Medicine*, 19(3), 989–996.

45. Ang, KC *et al.* (2007). Compressive properties and degradability of poly(ε-caprolatone)/hydroxyapatite composites under accelerated hydrolytic degradation. *Journal of Biomedical Materials Research — Part A*, 80(3), 655–660.

46. Wiria, FE *et al.* (2007). Poly-ε-caprolactone/hydroxyapatite for tissue engineering scaffold fabrication via selective laser sintering. *Acta Biomaterialia*, 3(1), 1–12.

47. Simpson, RL *et al.* (2008). Development of a 95/5 poly(L-lactide-co-glycolide)/hydroxylapatite and β-tricalcium phosphate scaffold as bone replacement material via selective laser sintering. *Journal of Biomedical Materials Research — Part B Applied Biomaterials*, 84(1), 17–25.

48. Wang, W, GF Caetano, W.-H Chiang, AL Braz, JJ Blaker, MAC Frade and PJDS Bartolo (2016). Morphological, mechanical and biological assessment of PCL/pristine graphene scaffolds for bone regeneration. *International Journal of Bioprinting*, 2(2), 95–105. doi: 10.18063/IJB.2016.02.009

49. Tan, JY, CK Chua and KF Leong (2010). Indirect fabrication of gelatin scaffolds using rapid prototyping technology. *Virtual and Physical Prototyping*, 5(1), 45–53.

50. Tan, JY, CK Chua and KF Leong (2012). Fabrication of channeled scaffolds with ordered array of micro-pores through microsphere leaching and indirect Rapid Prototyping technique. *Biomedical Microdevices*, 1–14.

51. Tan, JY, CK Chua and KF Leong (2013) Fabrication of channeled scaffolds with ordered array of micro-pores through microsphere leaching and indirect Rapid Prototyping technique. *Biomedical Microdevices*, 15(1), 83–96.

52. Yeong, WY *et al.* (2006). Indirect fabrication of collagen scaffold based on inkjet printing technique. *Rapid Prototyping Journal*, 12(4), 229–237.

53. Yeong, WY *et al.* (2007). Comparison of drying methods in the fabrication of collagen scaffold via indirect rapid prototyping. *Journal of Biomedical Materials Research — Part B Applied Biomaterials*, 82(1), 260–266.

54. Liu, MJJ *et al.* (2013). The development of silk fibroin scaffolds using an indirect rapid prototyping approach: Morphological analysis and cell growth monitoring by spectral-domain optical coherence tomography. *Medical Engineering and Physics*, 35(2), 253–262.

55. Tay, BCM *et al.* (2013). Monitoring cell proliferation in silk fibroin scaffolds using spectroscopic optical coherence tomography. *Microwave and Optical Technology Letters*, 55(11), 2587–2594.

56. Ramanath, HS *et al.* (2008). Melt flow behaviour of poly-ε-caprolactone in fused deposition modelling. *Journal of Materials Science: Materials in Medicine*, 19(7), 2541–2550.

57. Wiria, FE, KF Leong and CK Chua (2010). Modeling of powder particle heat transfer process in selective laser sintering for fabricating tissue engineering scaffolds. *Rapid Prototyping Journal*, 16(6), 400–410.

58. Wiria, FE *et al.* (2010). Selective laser sintering adaptation tools for cost effective fabrication of biomedical prototypes. *Rapid Prototyping Journal*, 16(2), 90–99.

59. Yeong, WY, Sudarmadji N, Yu HY, Chua CK, Leong KF, Venkatraman SS, Boey YC and Tan LP (2010). Porous polycaprolactone scaffold for cardiac tissue engineering fabricated by selective laser sintering. *Acta Biomaterialia*, 6(6), 2028–2034.

60. Sánchez-Salcedo, S, M Colilla, I Izquierdo-Barba and M Vallet-Regí (2016). Preventing bacterial adhesion on scaffolds for bone tissue engineering. *International Journal of Bioprinting*, 2(1), 20–34. doi: 10.18063/IJB.2016.01.008

61. Chua, CK and WY Yeong (2014). *Bioprinting: Principles and Applications*. Singapore: World Scientific.

62. Sundaramurthi, D, S Rauf and CAE Hauser (2016). 3D bioprinting technology for regenerative medicine applications. *International Journal of Bioprinting*, 2(2), 9–26. doi: 10.18063/IJB.2016.02.010

63. Tan, EYS and WY Yeong (2015). Concentric bioprinting of alginate-based tubular constructs using multi-nozzle extrusion-based technique. *International Journal of Bioprinting*, 1(1), 49–56. doi: 10.18063/IJB.2015.01.003

64. Tse, CCW, SS Ng, J Stringer, S MacNeil, JW Haycock and PJ Smith (2016). Utilising inkjet printed paraffin wax for cell patterning applications. *International Journal of Bioprinting*, 2(1), 35–44. doi: 10.18063/IJB.2016.01.001

65. Hariharan, K and G Arumaikkannu (2016). Structural, mechanical and in vitro studies on pulsed laser deposition of hydroxyapatite on additive manufactured polyamide substrate. *International Journal of Bioprinting*, 2(2), 85–94. doi: 10.18063/IJB.2016.02.008

66. Taidi, B, G Lebernede, L Koch, P Perre and B Chichkov (2016). Colony development of laser printed eukaryotic (yeast and microalga) microorganisms in co-culture. *International Journal of Bioprinting*, 2(2), 37–43. doi: 10.18063/IJB.2016.02.001

67. Arai, K, Y Tsukamoto, H Yoshida, H Sanae, TA Mir, S Sakai, T Yoshida, M Okabe, T Nikaido, M Taya and M Nakamura (2016). The development of cell-adhesive hydrogel for 3D printing. *International Journal of Bioprinting*, 2(2), 44–53. doi: 10.18063/IJB.2016.02.002

68. Li, H, S Liu and L Li (2016). Rheological study on 3D printability of alginate hydrogel and effect of graphene oxide. *International Journal of Bioprinting*, 2(2), 54–66. doi: 10.18063/IJB.2016.02.007

69. Mehrban, N, GZ Teoh and MA Birchall (2016). 3D bioprinting for tissue engineering: stem cells in hydrogels. *International Journal of Bioprinting*, 2(1), 6–19. doi: 10.18063/IJB.2016.01.006

70. Ng, WL, WY Yeong and MW Naing (2016). Polyelectrolyte gelatin-chitosan hydrogel optimized for 3D bioprinting in skin tissue engineering. *International Journal of Bioprinting*, 2(1), 53–62. doi: 10.18063/IJB.2016.01.009

71. Wang, S, JM Lee and WY Yeong (2015). Smart hydrogels for 3D bioprinting. *International Journal of Bioprinting*, 1(1), 3–14. doi: 10.18063/IJB.2015.01.005

72. Definition of bioprinting in English. http://www.oxforddictionaries.com/definition/english/bioprinting

73. Bhuthalingam, R, PQ Lim, SA Irvine, A Agrawal, PS Mhaisalkar, J An, CK Chua and S Venkatraman (2015). A novel 3D printing method for cell alignment and differentiation. *International Journal of Bioprinting*, 1(1), 57–65. doi: 10.18063/ijb.2015.01.008

74. An, J, CK Chua, T Yu, H Li and LP Tan (2013). Advanced nanobiomaterial strategies for the development of organized tissue engineering constructs. *Nanomedicine*, 8(4), 591–602. doi: 10.2217/nnm.13.46

75. An, J, JEM Teoh, R Suntornnond and CK Chua (2015). Design and 3D printing of scaffolds and tissues. *Engineering*, 1(2), 261–268. doi: 10.15302/J-ENG-2015061

76. An, J, CK Chua and V Mironov (2016). A Perspective on 4D bioprinting. *International Journal of Bioprinting*, 2(1), 3–5. doi: 10.18063/ijb.2016.01.003

77. Wang, M, J He, Y Liu, M Li, D Li and Z Jin (2015). The trend towards in vivo bioprinting. *International Journal of Bioprinting*, 1(1), 15–26. doi: 10.18063/IJB.2015.01.001

78. Mironov, V, T Boland, T Trusk, G Forgacs and RR Markwald (2003). Organ printing: computer-aided jet-based 3D tissue engineering. *Trends in Biotechnology,* 21(4), 157–161. doi: 10.1016/S0167-7799(03)00033-7

79. Koudan, EV, EA Bulanova, FDAS Pereira, VA Parfenov, VA Kasyanov, YD Hesuani and VA Mironov (2016). Patterning of tissue spheroids bio-fabricated from human fibroblasts on the surface of electrospun polyurethane matrix using 3D bioprinter. *International Journal of Bioprinting,* 2(1), 45–52. doi: 10.18063/IJB.2016.01.007

80. Nakamura, M, TA Mir, K Arai, S Ito, T Yoshida, S Iwanaga, H Kitano, C Obara and T Nikaido (2015). Bioprinting with pre-cultured cellular constructs towards tissue engineering of hierarchical tissues. *International Journal of Bioprinting,* 1(1), 39–48. doi: 10.18063/IJB.2015.01.007

81. Lam, CR, HK Wong, S Nai, CK Chua, NS Tan and LP Tan (2014). A 3D biomimetic model of tissue stiffness interface for cancer drug testing. *Molecular Pharmaceutics,* 11(7), 2016–2021. doi: 10.1021/mp500059q

82. Knowlton, S, A Joshi, B Yenilmez, IT Ozbolat, CK Chua, A Khademhosseini and S Tasoglu (2016). Advancing cancer research using bioprinting for tumor-on-a-chip platforms. *International Journal of Bioprinting,* 2(2), 3–8. doi: 10.18063/ijb.2016.02.003

83. Bibb, R, N Nottrodt and A Gillner (2016). Artificial vascularized scaffolds for 3D-tissue regeneration — a report of the ArtiVasc 3D Project. *International Journal of Bioprinting,* 2(1), 93–102. doi: 10.18063/IJB.2016.01.004

84. Lee, JM., SL Sing, EYS Tan and WY Yeong (2016). Bioprinting in cardiovascular tissue engineering: a review. *International Journal of Bioprinting,* 2(2), 27–36. doi: 10.18063/IJB.2016.02.006

85. Liu, L and X Wang (2015). Creation of a vascular system for organ manufacturing. *International Journal of Bioprinting,* 1(1), 77–86. doi: 10.18063/IJB.2015.01.009

86. Leong, KF *et al.* (2001). Fabrication of porous polymeric matrix drug delivery devices using the selective laser sintering technique. *Proc. of the Institution of Mechanical Engineers, Part H: Journal of Engineering in Medicine,* 215(2), 191–201.

87. Low, KH *et al.* (2001). Characterization of SLS parts for drug delivery devices. *Rapid Prototyping Journal,* 7(5), 262–267.

88. Cheah, CM *et al.* (2002). Characterization of microfeatures in selective laser sintered drug delivery devices. *Proc. of the Institution of Mechanical Engineers, Part H: Journal of Engineering in Medicine*, 216(6), 369–383.

89. Leong, KF *et al.* (2006). Building porous biopolymeric microstructures for controlled drug delivery devices using selective laser sintering. *International Journal of Advanced Manufacturing Technology*, 31(5–6), 483–489.

90. Leong, KF *et al.* (2007). Characterization of a poly-ε-caprolactone polymeric drug delivery device built by selective laser sintering. *Bio-Medical Materials and Engineering*, 17(3), 147–157.

91. Liew, CL *et al.* (2001). Dual material rapid prototyping techniques for the development of biomedical devices. Part 1: Space creation. *The International Journal of Advanced Manufacturing Technology*, 18(10), 717–723.

92. Liew, CL *et al.* (2002). Dual material rapid prototyping techniques for the development of biomedical devices. Part 2: Secondary powder deposition. *The International Journal of Advanced Manufacturing Technology,* 19(9), 679–687.

93. Total Contact. (2007). http://www.totalcontact.com

94. Materialise NV. (2007). http://www.cadinfo.net/editorial/ honeywell.htm

95. Cesarani, F, MC Martina, RR Grilletto, R Boano, AM Roveri, V Capussotto, A Giuliano, M Celia and G Gandini (2004). Facial reconstruction of a wrapped Egyptian mummy using MDCT. *American Journal of Roentgenology*, 183(3), 755–758.

96. Crockett, RS and R Zick (2000). Forensic applications of solid freeform fabrication. In *Proceedings for the Solid Freeform Fabrication Symposium*, pp. 549–554.

97. Shimizu, TS, N Le Novère, MD Levin, AJ Beavil, BJ Sutton and D Bray (2000). Molecular model of a lattice of signalling proteins involved in bacterial chemotaxis. *Nature Cell Biology*, 2(11), 792–796.

98. Laub, M, M Chatzinikolaidou, H Rumpf and HP Jennissen (2002). Modelling of protein-protein interactions of bone morphogenetic protein-2 (BMP-2) by 3D-rapid prototyping. *Materialwissenschaft und Werkstofftechnik*, 33, 729–737.

99. Mannoor, MS *et al.* (2013). 3D Printed Bionic Ears. *Nano Letters*, 13(6), 2634–2639.

100. Young, S (2013). *Cyborg Parts.*
http://www.technologyreview.com/demo/517991/cyborg-parts/

101. EnvisionTEC.
http://envisiontec.com/case_studies/highland-dental-laboratory-produces-crowns-bridges-3d-printer/

102. Collins, K. (October 2013). *Blizzident 3D-printed toothbrush cleans your gnashers in six seconds.* Wired.co.uk.
http://www.wired.co.uk/news/archive/2013-10/01/blizzident

103. Grynol, B (2013). *Disruptive Manufacturing: The effects of 3D printing.* Deloitte.
http://www.deloitte.com/assets/DcomCanada/Local%20Assets/Documents/Insights/Innovative_Thinking/2013/ca_en_insights_disruptive_manufacturing_102813.pdf

Problems

1. List several possible applications for AM in medical and biomedical engineering.

2. Name several advantages and disadvantages of applying AM to the field of medicine and biomedical engineering.

3. List some possible materials for use with AM in relation to in-vivo applications.

4. Discuss how AM can create value for surgical procedures relating to the separation of conjoined twins joined at the head.

5. How can AM models be useful to surgeons before and during the operating procedure to remove a tumor from the cranium?

6. Why and how is AM important when producing hip implants for non-standard sized patients requiring hip replacement?

7. Respiratory stents, such as the buccopharyngeal stent, have been used on babies with congenital buccopharyngeal defects. Explain how AM can be used to support the child with such a stent until he is old enough for major reconstructive surgery.

8. Discuss how AM can be used to support organ replacement by tissue engineering. Discuss what materials should be considered and why.

9. Explain the challenges of building an AM system for tissue engineering applications.

10. In what ways can an AM model assist in the design and laboratory testing of a substitution replacement mechanical heart valve? Discuss the AM selection considerations for such an application.

11. How can AM prove useful in forensic science applications? Compare the use of current techniques versus AM for such applications.

12. Being able to visualize scientific concepts in three-dimensions can be quite a challenge when it comes to protein-protein interaction research. How can AM create value in research or teaching laboratories for such an application?

13. Describe how a 3D printer is used to manufacture a bionic ear. What are the advantages of using 3D printing in this application?

14. Give two examples of the applications of AM in dentistry.

CHAPTER 9

BENCHMARKING AND THE FUTURE
OF ADDITIVE MANUFACTURING

9.1 Technical Evaluation Through Benchmarking

The execution of a benchmark test is a traditional practice, necessary for all kinds of highly productive and expensive equipment, such as computer-aided design (CAD)/computer-aided manufacturing (CAM) workstation, computer numeric control (CNC) machining centre, and so on. Wherever a relatively broad spread of possibilities is offered for specific users' requirements, the execution of a benchmark test is absolutely necessary. The dynamic development and increased range of commercially available additive manufacturing (AM) systems on offer (currently more than 30 different types of equipment worldwide, partly in different types of equipment, partly in different sizes), mean objective decision making is essential. In analysing the benchmark test piece, some tests have to be conducted including visual inspection and dimension measurement.

9.1.1 *Reasons for benchmarking*

Generally, benchmarking serves the following purposes:

(1) It is a valuable tool for evaluating strengths and weaknesses of the systems tested. Vendors have to produce the benchmark models in response to requests from potential buyers. In doing so, they will not have the choice to demonstrate what they want, but have to show what is requested. Consequently, vendors cannot hide the limitations of the system. It is also a rigorous and therefore more

revealing means of testing so that the potential buyer can verify the claims of the vendor.

(2) Since the benchmark model is specifically designed for the potential buyer, it can be custom-made to its own requirements and needs. For example, in the case of a company that makes parts which frequently have very thin walls, then the ability of the system to produce accurately built thin walls can be tested, measured and verified.

(3) Benchmarking has also become a means of helping various departments within a company comprehend what the AM system can do for them. This is vital in the context of concurrent engineering, whereby designers, analysts, manufacturing engineers work on the product concurrently. Today, an AM system's application areas extend beyond design models, to functional models and manufacturing models.

(4) Sometimes, a benchmark test may also help to identify applications for an AM system, which had not previously been considered. Although this is not really a primary motivational factor for benchmarking, it is nevertheless a side benefit.

9.1.2 *Benchmarking methodology*

There are four steps in the proposed benchmarking methodology:

(1) Deciding on the benchmarking model type.
(2) Deciding on the measurements.
(3) Recording time and measurements, tabulating and plotting the results.
(4) Analysing and comparing results.

9.1.2.1 *Deciding on the benchmarking model type*

In general, additive manufacture benchmarking models can be categorised according to the following types:

Table 9.1: Part structure classification scheme of 10 part classes.

Part Class Number	Part Structure
Part Class 1	Compact parts
Part Class 2	Hollow parts
Part Class 3	Box type
Part Class 4	Tubes
Part Class 5	Blades
Part Class 6	Ribs, profiles
Part Class 7	Cover type
Part Class 8	Flat parts
Part Class 9	Irregular parts
Part Class 10	Mechanisms

Figure 9.1: Part structure classification scheme for AM systems.[1]

(1) *Typical company products.* This is probably the most common type since the company needs to confirm how well its products can be prototyped and whether its requirements can be fulfilled. The company is also in the best position to comment on the results since it has intimate product knowledge. Examples are turbine blades, jewellery, cellular phones, and so on.

(2) *Part classification.* According to the thesis of Wall,[1] a classification scheme based on general part structures is applicable for AM parts.

This classification scheme, based on part structure, has 10 part classes, as seen in Table 9.1.

Fig. 9.1 shows some examples of such a classification scheme. The part sizes and part structures are mostly related to shrinkage, distortion and curling effects.

(3) *Complex and hybrid parts.* Alternatively, determined complex parts can be designed, aiming to test the performance of available systems in specific aspects. Furthermore, this category can include a hybrid combining types 1 and 2 above.

9.1.2.2 *Deciding on the measurements*

In deciding the measurements, it must be stressed that, as far as possible, the benchmark model should:

(1) be relatively simple and designed with low expenditure,
(2) not utilise too much material and
(3) allow simple measuring devices to determine the measurements.

In general, two types of measurements can be taken, namely main (large) measurements and detailed (small) measurements.

9.1.2.3 *Recording time and measurements, tabulating and plotting the results*

The measurement results are based on the deviations of the part built from the CAD model. These deviations of both the main and detailed measurements are tabulated for each of the AM systems. Subsequently, the results can be plotted to graphically present the performance of the systems in a single diagram. Thus, for the main measurements, one can visually compare the systems' performance using the main measurement diagram and similarly, using the detailed measurement diagram.

The time results are based on three components — data preparation, building time and post-processing. The total time is based on the addition of the three components. A table of all four-time results can be tabulated with each AM system alongside another.

9.1.2.4 *Analysing and comparing results*

In a general analysis, the deviations of all systems may or may not be in an acceptable range. The evaluator of the systems is usually decided by this acceptable range. The time component by itself gives one an idea of the length required for a task and directly affects the cost factor. Therefore, the time data can become useful for a full economic justification and cost analysis.

In comparison, one can determine, based on the time and measurement results, the strengths and weaknesses of the systems. To arrive at a conclusion, the evaluator must only consider the benchmarking results as *a component* of the overall evaluation study. The benchmarking results should never be taken as the only deciding factor of an evaluation study.

Finally, the above approach ignores the human aspects. For example, the skills, expertise and experience of the vendors' operators are not accounted for in the benchmark.

9.1.3 *Case study*

9.1.3.1 *The button tree display*

With the compliments of Thomson Multimedia in Singapore, the results of a benchmarking study involving five machines are made available here. The test piece is a button tree display that is mounted in between the front cabinet of a hi-fi set and printed circuit board (PCB), as shown in Fig. 9.2. The button tree display has a frame of length 128 mm and width 27 mm. It consists of three round buttons joined to the frame by

0.6 mm hinges. The buttons have "legs" that contact the tact switch on the PCB when it is depressed. There are also locating pins for location and light-emitting diode (LED) holders on the frame. Catches are made from the side of the frame so that the button tree display can hold down firmly to the PCB. The five different test pieces are made from the principles A, B, C, D and E. A photograph of each test piece is shown in Figs. 9.3–9.7, respectively.

The measurements taken are linear, radial and angular dimensions. Coordinate measuring machine (CMM), profile projector and vernier calliper are used for the measurements. When choosing the type of dimensions to measure, a few criteria are considered:

(1) The overall dimensions are to be included.
(2) The important dimensions that will affect the operation of the button tree.
(3) There must be a variety of dimensions.
(4) There must be sufficient main and detailed dimensions to plot the graph and give an indication of which method is superior.

Figure 9.2: Button tree display is mounted on the front cabinet of a hi-fi.

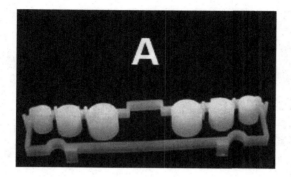

Figure 9.3: Benchmark test piece made from principle A.

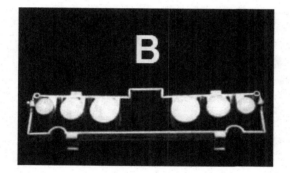

Figure 9.4: Benchmark test piece made from principle B.

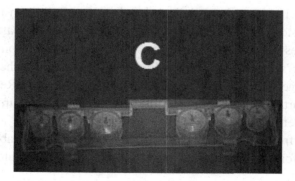

Figure 9.5: Benchmark test piece made from principle C.

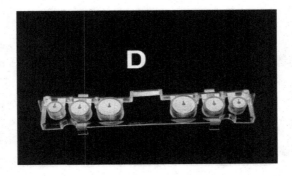

Figure 9.6: Benchmark test piece made from principle D.

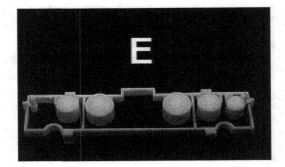

Figure 9.7: Benchmark test piece made from principle E.

Figs. 9.8–9.10 show the technical drawings for the button tree display. The dimensions of the test piece are divided into three parts. The first part includes dimensions taken from the frame, locating pins and LED holder. The second part has the dimensions taken from the button set, and the dimensions of the catch fall into the third part. Related to the different parts, the results are sub-divided into main measurements (>10 mm) and detailed measurement (= 10 mm). For both the readings, the deviations from the actual reading are computed and tabulated. The deviations for both the measurements are plotted and compared.

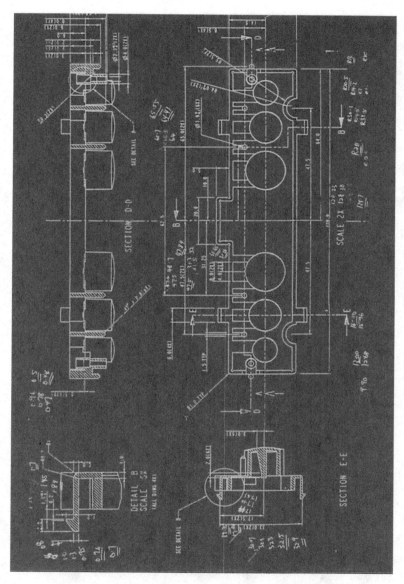

Figure 9.8: Front and plan views of the Button Tree Display.

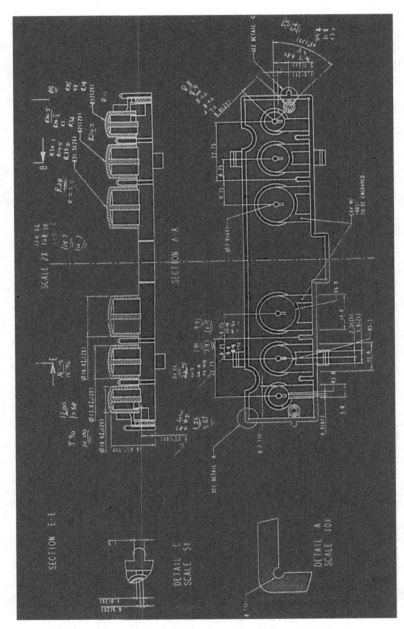

Figure 9.9: Sectional view of the Button Tree Display.

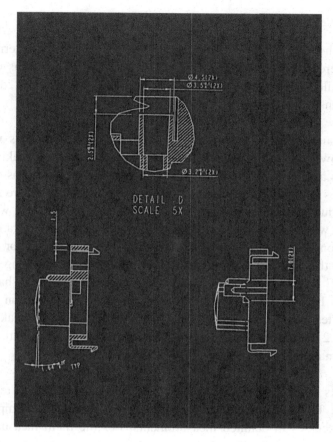

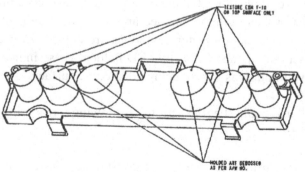

Figure 9.10: (Top) Sectional and (bottom) isometric views of the Button Tree Display.

9.1.3.2   *Results of the measurements*

The graphs are plotted to facilitate a detailed evaluation concerning the five different techniques, namely A, B, C, D and E, in terms of main and detailed measurements and their deviations from the designed dimensions.

From visual inspections, neither the thin rib nor the wall is missing. This shows that all the five processes are capable of making wall thickness as thin as 0.5 mm. However, one of the locating pins of the B test pieces is tilted and the catches of the A and E test pieces are missing. This is due to the catches being too weak and coming off when the supports were removed. SLA objects are built on supports rather than directly on the elevator platform. Supports are used to anchor the part firmly to the platform and prevent distortion during the building process. The C, E and D test pieces came with supports, thus they had to be removed before any measurements were taken. The button sets of the five test pieces are slanted due to its weight and the weak hinges. Tables 9.2 and 9.3 list the measurements taken and their deviations from the nominal values.

The plots illustrated in Figs. 9.11 and 9.12 show that A test piece achieved better measurements or fewer deviations as compared to other processes in detailed measurements. For the main measurements, E attained fewer deviations. On the other hand, C test piece produced the higher deviations for both the detailed and main measurements. From all the measurements taken, greater deviations are observed for the radii of the button set and the angles of the LED holder. These indicate that AM systems have limitations in producing good curve surfaces.

The deviations could be due to the following errors in taking the measurements:

(1) The poor surface finishes of the test pieces result in inaccurate dimensions.

(2) The test pieces are not strong as they became deformed under pressure.

(3) The support of the C test piece was not properly removed.

Table 9.2: Main measurements (>10 mm) of the five benchmark test pieces.

Main Measurements					
Drawing Dimensions (mm)	**Measured Dimensions (mm) and Deviations (mm) in *italics***				
	A	**B**	**C**	**D**	**E**
128.0	128.3	129.1	128.2	128.7	128.7
	+0.3	*+1.1*	*+0.2*	*+0.7*	*+0.7*
27.0	27.1	27.3	27.1	27.1	26.8
Frame,	*+0.1*	*+0.3*	*+0.1*	*+0.1*	*-0.2*
Location	65.0	65.7	65.5	64	64.4	65.6
Pins and		*+0.7*	*+0.5*	*-1.0*	*-0.6*	*+0.6*
LED Hold	47.5	47.6	47.5	48.7	47.9	48.7
	+0.1	*0.0*	*+1.2*	*+0.4*	*+1.2*
31.25	31.3	31.5	32	31.7	31.2
	+0.05	*+0.25*	*+0.75*	*+0.4*	*-0.05*
R31.5	30.1	14.5	33.5	28.0	29.5
	-1.4	*-17.0*	*+2.0*	*-3.5*	*-2.0*
R21.0	14.2	7.0	32.0	20.5	20.5
	-6.8	*-14.0*	*+11.0*	*-0.5*	*-0.5*
R12.0	15.0	8.0	14.0	13.0	11.0
Button-Tree	*+3.0*	*-4.0*	*+2.0*	*+1.0*	*-1.0*
16.0	16.1	16.1	15.9	16.0	16.0
	+0.1	*+0.1*	*-0.1*	*0.0*	*+0.0*
13.0	13.0	13.1	12.9	13.0	12.7
	+0.0	*+0.1*	*-0.1*	*0.0*	*-0.3*
22.75	23.0	23.8	24.4	22.45	22.3
	+0.25	*+1.05*	*+1.65*	*-0.3*	*-0.45*
33.0	32.3	33.1	32.3	32.5	32.9
Catch	*-0.7*	*+0.1*	*-0.7*	*-0.5*	*-0.1*
17.5	17.4	17.0	17.6	17.8	17.4
	-0.1	*-0.5*	*+0.1*	*+0.3*	*-0.1*

Table 9.3: Detailed measurements (= 10 mm) of the five benchmark test pieces.

Detailed Measurements						
Drawing Dimensions (mm)		**Measured Dimensions (mm) and Deviations (mm) in *italics***				
		A	**B**	**C**	**D**	**E**
Location Pins	8.0	7.2	8.2	8.0	8.1	8.8
		−0.8	*+0.2*	*+0.0*	*+0.1*	*+0.8*
	5.7	4.7	6.8	6.5	7.6	7.4
		−1.0	*−1.1*	*−0.8*	*−1.9*	*−1.7*
	45°	39.4	31.8	37	56	54
		−5.6	*−13.2*	*−8.0*	*+11.0*	*+9.0*
	8.25	8.8	8.0	8.4	8.0	7.9
		+0.55	*−0.25*	*+0.15*	*−0.25*	*−0.35*
Button-Tree	9.75	10.1	10.2	10.7	10.1	10.0
		+0.35	*+0.45*	*+0.95*	*+0.35*	*+0.25*
	10.0	10.1	10.1	9.9	10.0	9.9
		+0.1	*+0.1*	*−0.1*	*0.0*	*−0.1*
	5.2	4.9	4.8	4.5	3.6	3.7
		−0.3	*−0.4*	*−0.7*	*−1.6*	*−1.5*
	0.5	0.5	1.0	0.8	0.5	0.4
		0.0	*+0.5*	*+0.3*	*0.0*	*−0.1*
Catch	1.6	1.5	1.1	1.4	1.7	1.9
		−0.1	*−0.5*	*−0.2*	*+0.1*	*+0.3*
	0.8	1.0	1.0	1.3	1.0	0.7
		+0.2	*+0.2*	*+0.5*	*+0.2*	*−0.1*

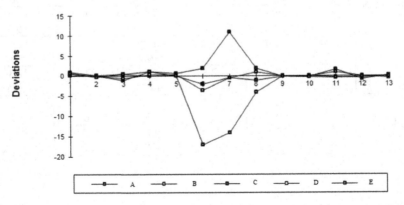

Figure 9.11: Deviation of main measurements (>10 mm) from the nominal values for the five benchmark test pieces.

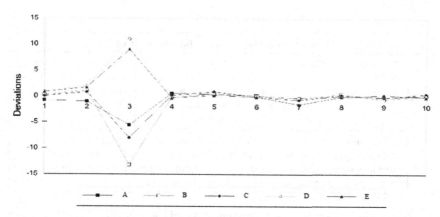

Figure 9.12: Deviation of detailed measurements (= 10 mm) from the nominal values for the five benchmark test pieces.

9.1.4 *Other benchmarking case studies*

Other than the benchmarking performed by Thomson Multimedia, there are also other benchmarking case studies done by various researchers. Tables 9.4(a)–(e) list a few from the year 2004 to 2013.[2–12]

Table 9.4(a): Benchmarking study published in 2013.

Benchmarking study published in 2013		
1.	M. Fahad and N. Hopkinson	Evaluation of parts produced by a novel AM process	Applied Mechanics and Materials, Volume 315 (2013) 63–67
	This benchmark analyses the flatness of parts produced on High-Speed Sintering (HSS), one of the AM processes being developed at Loughborough University. The designed benchmark part comprised various features such as cubes, holes, cylinders and cones on a flat base.		

Table 9.4(b): Benchmarking study published in 2012.

Benchmarking study published in 2012			
2.	E. Atzeni, L. Iuliano, P. Minetola and A. Salmi	Proposal of an innovative benchmark for accuracy evaluation of dental crown manufacturing	Computers in Biology and Medicine, Vol. 42, Issue 5, (2012) 548–555
	This innovative benchmark represents a dental arch with features relating to different types of prepared teeth. It includes tooth orientation and oblique surfaces similar to those of real prepared teeth. The evaluation procedure proves that the scan data can be used as a reference model for crown restoration design.		

Table 9.4(c): Benchmarking study published in 2011.

Benchmarking study published in 2012			
3.	W. M. Johnson, M. Rowell, B. Deason, and M. Eubanks	Benchmarking evaluation of open source fused deposition modeling (FDM) system	Proceeding of the 22nd Annual International Solid Freeform Fabrication Symposium (pp. 197–211).
	This paper reviews previous benchmarking models and presents the development of a new benchmarking model and its application in the evaluation of an open source AM system based on FDM. The proposed benchmarking model includes various geometric features to evaluate the AM system in terms of dimensional accuracy, thermal warpage, staircase effect and geometric and dimensional tolerances.		

Table 9.4(d): Benchmarking study published in 2007.

Benchmarking study published in 2007			
4.	B. Vandenbroucke and J.P. Kruth	Selective laser melting (SLM) of biocompatible metals for rapid manufacturing of medical parts	Rapid Prototyping Journal, Vol. 13 Issue: 4, pp.196–203
	This paper seeks to investigate the possibility of producing medical or dental parts by SLM. The benchmark considers the mechanical and chemical properties and geometrical feasibility, including process accuracy and surface roughness. By developing a procedure to fabricate frameworks for complex dental prostheses, the potential of SLM as a medical manufacturing technique has been proved.		

Table 9.4(e): Benchmarking studies published in 2006.

Benchmarking studies published in 2006		
5.	K. Abdel Ghany and S.F. Moustafa	Comparison between the products of four RPM systems for metals	Rapid Prototyping Journal 12/2 (2006) 86–94
	This work evaluates and compares the quality of four identical benchmarks fabricated from different metallic powders by using four recently developed RPM systems for metals. The evaluation considers benchmark geometry, dimensional precision, material type, product strength and hardness, surface quality, building speed, materials, operation and running cost.		
6.	D. Dimitrov, W. van Wijck, K. Schreve and N. de Beer	Investigating the achievable accuracy of three-dimensional printing (3DP)	Rapid Prototyping Journal 12/1 (2006) 42–52
	This paper deals with current research towards the building of a full capability profile — accuracy, surface roughness, strength, elongation, build time and cost — of this important process.		
7.	M. Mahesh, Y.S. Wong, J.Y.H. Fuh and H.T. Loh	A Six-Sigma approach for Benchmarking of RP&M processes	International Journal of Advanced Manufacturing Technology (2006) 31: 374–387
	This paper presents a methodology of using six-sigma quality tools for the benchmarking of rapid prototyping & manufacturing (RP&M) processes. It involves the fabrication of a geometric benchmark part and a methodology to control and identify the best performance of the process to reduce variability in the fabricated parts.		

Table 9.4(f): Benchmarking studies published in 2005.

Benchmarking studies published in 2005		
8.	Todd Grimm	3D Printer Dimensional Accuracy Benchmark	T.A. Grimm & Associates, Inc.
	This benchmark analyses and quantifies the dimensional accuracy available from the Dimension® SST, InVision ™ SR and ZPrinter® 310. This report illustrates the accuracy of each system with reverse engineering colour maps and comparative charts.		

Table 9.4(f): *(Continued)* Benchmarking studies published in 2005.

9.	Vito R. Gervasi, Adam Schneider and Joshua Rocholl	Geometry and procedure for benchmarking SFF and Hybrid fabrication process resolution	Rapid Prototyping Journal 11/1 (2005) 4–8
	This paper shares with the solid freeform fabrication community a new procedure and benchmark geometries for evaluating SFF process capabilities. The procedure evaluates the range capability of various SFF and SFF-based hybrid processes in producing rod and hole elements.		

Table 9.4(g): Benchmarking studies published in 2004.

Benchmarking studies published in 2004			
10.	M. Mahesh, Y.S. Wong, J.Y.H. Fuh and H.T. Loh	Benchmarking for comparative evaluation of AM systems and processes	Rapid Prototyping Journal Vol 10, number 2, 2004 (pp. 123–135)
	This paper presents issues on AM benchmarking and aims to identify factors affecting the definition, fabrication, measurements and analysis of benchmark parts.		
11.	K.W. Dalgarno and R.D Goodridge	Compression Testing of layer manufactured metal parts: the RAPTIA compression benchmark	Rapid Prototyping Journal, Vol 10, number 4, 2004 (pp 261–264)
	This paper reports the results of a compression test benchmarking study carried out to investigate the mechanical properties of layer manufactured metal components in order to assess their suitability for load-bearing applications. Compression tests were carried out on the DTM LaserForm St-100 material, ARCAM processed H13 tool steel, EOS DirectSteel (50 µm), and ProMetal material.		

9.2 Industrial Growth

9.2.1 *Industrial growth of AM*

The AM industry, formerly known as rapid prototyping, has enjoyed tremendous growth since the first system was introduced in 1988. The industry has continued to grow over the past 6 years after a decline in

2009. The compound annual growth rate (CAGR) of worldwide revenues produced by all products and services over the past 27 years is an impressive 26.2%. The CAGR for the past 3 years (2013–2015) is 31.5%.

The AM industry has grown in the double digits for 19 of its 28 years. It continues to offer great potential, especially in custom and short-run part production. AM system manufacturers and service providers are increasingly offering solutions for the production of parts that go into final products. However, this market segment comes with much higher quality standards than those associated with modelling and prototyping applications. As it continues to develop, the demand for production parts from AM is expected to drive annual revenues to much higher levels.

In 2015, the AM industry, consisting of all AM products and services worldwide, grew 25.9% (CAGR) to $5.165 billion. The growth in 2015 compares to 35.2% growth in 2014 when the industry reached $4.103 billion. Industry-wide growth in 2013, 2012 and 2011 was 33.4%, 32.7% and 29.4%, respectively.[13] These estimates consist of revenues generated in the primary AM market which include all products and services directly associated with AM processes worldwide.

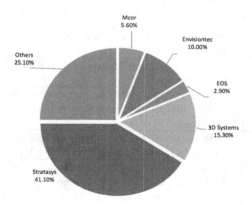

Figure 9.13: Unit sales market share estimates among manufacturers of industrial AM systems worldwide.
Adapted from Wohler's Report 2016.

Currently, the two giants in the AM industry, 3D Systems and Stratasys, are both undergoing rapid growth. According to Wohlers Report 2016,[13] Stratasys sold an estimated 5166 industrial AM systems in 2015, down 22.5% from an estimated 6665 in 2014. This estimate includes FDM and PolyJet systems, as well as machines sold by Solidscape (a Stratasys company), but excludes MakerBot sales. The following chart (see Fig. 9.13) shows the 2015 unit sales market share estimates among the manufacturers of industrial AM systems worldwide. Stratasys' share fell from 51.9% in 2014 to 41.1% in 2015. However, the company still leads in industrial system unit sales for the 14th consecutive year.

The significant milestones of AM from 1998 to 2014 are captured in Table 9.5.

Table 9.5: Significant milestones of AM.[14]

Time	Event
1988	3D Systems was founded by Charles Hull. The technique was named Stereolithography (SLA) and a patent was obtained
1988	3D Systems developed SLA-250, the first version of 3D printers made available to the public
1988	Fused Deposition Modelling (FDM) was invented by Scott Crump
1989	Stratasys was founded by Scott Crump
1989	EOS was founded
1991	The Laminated Object Manufacturing (LOM) system was sold by Helisys
1992	The Selective Laser Sintering (SLS) system was sold by DTM to 3D Systems
1993	An inkjet-based machine, which can produce outstanding surface finish with relatively low speed, was founded by Solidscape
1993	3 Dimensional Printing (3DP) techniques were patented by Massachusetts Institute of Technology (MIT)
1995	Z Corporation obtained the license to use 3DP techniques from MIT to develop its own 3D printers
1996	"Genisys", the legacy FDM product, was introduced by Stratasys
1996	The term "3D Printer" was first used to refer to all the AM machines
1997	EOS' SLA business was sold to 3D Systems

Table 9.5: (*Continued*) Significant milestones of AM.[14]

Time	Event
1997	The Electron Beam Melting (EBM) technique was used by Arcam AB to produce solid metal parts
2000	SLM technology was introduced to the market
2000	Concept Laser GmbH was founded to optimize the SLM process
2003	Selective Deposition Lamination (SDL) technology was invented by Dr. Conor MacCormack and Fintan MacCormack
2005	The first high resolution 3D printer was released by Z Corporation
2005	SLM started at the Fraunhofer Institute for Laser Technology in Germany
2006	SLM Solutions became the first company to process aluminium and titanium on SLM machines.
2008	Connex500™, which manufactures models with several different materials at the same time, was invented by Objet.
2010	The first 3D printed car, Urbee, was produced using Stratasys' 3D printers
2010	First fully bioprinted blood vessels were made by Organovo, Inc., a regenerative medicine company.
2011	3D food printer underwent development at Cornell University
2011	The travel speed of 3D printer was raised to 350 mm/second by Ultimaker, a Dutch 3D printer manufacturer
2011	First 3D printed bikini was manufactured by Shapeways and Continuum Fashion
2011	First 3D chocolate printer was invented by the University of Exeter, the University of Brunel and application developer Delcam in the UK
2011	First 3D printed aircraft was made by researchers at the University of Southampton
2011	RegenHU Ltd, which specialises in making bio-printers and producing 3D organomimetic models for tissue engineering, was founded
2012	ProJet 7000 SLA machine was released by 3D Systems
2012	3Z line of high-precision wax 3D printers and 3ZSupport were released by Solidscape
2013	Z Corporation was acquired by 3D Systems
2013	The first 3D printed gun, called the "Liberator", is created by Defense Distributed. The company was forced to take down the blueprint for the gun from its website by the Department of State.
2014	3D printing had its own TechZone at the 2014 International CES
2014	DMG MORI introduced its hybrid system LASERTEC 65 3D

9.3 Future Trends

As the whole AM industry moves forward, AM-driven activities will continue to grow. Compared to several years ago, more significant trends have appeared. Among them are the proliferation of low-cost 3D printers, reduction of cost of 3D printing due to expiring of patents, increase in metal 3D printing and AM in rapid manufacturing and tooling (RM&T). Continuous improvements in the area of speed and quality, ease of use and post-processing, standardisation of systems, new materials, home use and growing number of applications have also been observed.

9.3.1 *Low-cost office and desktop 3D printers*

Office and desktop 3D printers may have resulted from the further development of concept modellers, the breed of AM system (especially the development of FDM technology) geared towards producing prototypes for design reviews instead of physical testing or fully functional parts. This class of AM system is characterised by its higher speed, lower cost and weaker accuracy and resolution compared to the higher-end class. It has been found that since the concept modeller is designed to operate in design offices, not at workshops, they feature clean and safe operations. These aspects have exceptional market appeal and have now become one of the most important aspects in choosing an AM system.

Over time, the problems with accuracy, surface finish and material properties of 3D printers have been gradually addressed. Although current 3D printers are still of a different class from higher-end systems, the gap is narrowing. With the quality of 3D printers increasing, the system may capture more and more market share from high-end systems in the future. An example of a high-resolution desktop machine is Solidscape's 3Z series printers. They offer a layer thickness of only 0.0254 mm and are even used for creating investment casting patterns.

The growth of office and desktop 3D printers is also accompanied by the lowering of cost to purchase and maintain such systems. While in 2003 such a system may have cost around $50,000, in 2006, the price was already reduced to around $20,000. Examples of machines of this price class are the Stratasys Dimension and 3D Systems' LD 3D Printer.

In 2014, 3D Systems began offering personal 3D Printers such as Cube® 3 and ProJet® 1500, which are much more affordable and start from only $1500.[15] These machines can produce prototypes with a minimum layer height of 0.1 mm.[16] Meanwhile, Stratasys has also revealed its own desktop 3D printer called Moji, which can produce finely resolved models in nine colours. Its cost is around $9900.[17]

The increase in the number of office and desktop 3D printers produced and purchased is an indication that the market is enlarging to encompass smaller companies that have neither workshop-class facilities nor the funds to purchase and maintain high-end AM systems. With the reduction in the price of 3D printers, it will become more affordable for educational institutions below university level, such as polytechnics, colleges and schools, to purchase 3D printers.

9.3.2 *Reduction in cost of 3D printing due to expiring of patents*

The cost of printing will continue to drop while the quality of 3D printers continues to improve. These developments can be attributed to advanced 3D printing technologies becoming more accessible due to the expiration of key patents on pre-existing industrial processes. Many of these patents were issued before 2000 and are reaching the end of their lifespan. Hence these processes will not be monopolised by the original pioneers of the 3D printing industry.

9.3.2.1 *Liquid-based AM systems*

Among all the liquid-based AM systems, SLA printing process is considered to be the best existing desktop printing process when it comes to creating highly detailed precision parts. The SLA was patented by Charles (Chuck) W. Hull in 1986[18] and it expired in 2014. This has led to a new generation of AM technologies that has paved the way for more accessible 3D printing. For example, Carbon invented a faster process based on SLA — continuous liquid interface production (CLIP), which works by successively printing thin layers of an object using an ultraviolet laser focused on a vat of liquid resin. Besides, Formlabs is a pioneer in bringing SLA 3D printing to the desktop at an accessible price point.

9.3.2.2 *Powder-based AM systems*

The SLS powder-based printing process was developed and subsequently patented by Dr. Carl Deckard and his academic adviser, Dr. Jow Beaman in 1984. The patent was later acquired by 3D Systems and it expired in 2014. Similar to what happened soon after SLA 3D patents dried up, this has resulted in a rise of new 3D printer manufacturers aimed at bringing this expensive industrial printing process onto the desktops of a wide amount of users.

The costs of metal 3D industrial printers and the maintenance are relatively high. There are two major metal-based AM systems — SLM and Direct Metal Laser Sintering (DMLS). However, a foundation patent for SLM held by Germany's Fraunhofer Institute for Laser Technology will be expiring in December 2016. This expiration is expected to bring with it a new group of manufacturers that will drive cost down dramatically.

9.3.2.3 *Solid-based AM system*

The FDM printing process patent expired in 2009. Following that, prices for FDM printers dropped from over $10,000 to less than $1000 and a new crop of consumer-friendly 3D printer manufacturers, like MakerBot and Ultimaker, made 3D printing more accessible.

With the steady growth in consumer 3D printer sales — nearly 200,000 units of desktop 3D printers (priced $5000 or below) were sold in 2015 alone — and the industry itself expecting to grow from generating approximately \$4.1 billion in 2015 to as much as \$16.2 billion within the next 4 years; the crumbling of monopolies and price disruptions, desktop 3D printing is moving forward beyond the hype and industrial quality desktop printing will be more accessible.

9.3.3 *Increase in demand for metal printing*

As its name implies, metal-based AM systems offer output material of metal; as such, they have excellent potential to produce ready-to-use prototypes of metal parts. Also called direct metal technologies,[19] these systems possess an absolute advantage in delivering prototypes of metal parts with closer material properties with the final product than resin-based or plastic-based AM systems while keeping the speed above regular CNC machining. Once the technology is perfected, it is possible for the systems to become rapid manufacturing systems, replacing traditional metal manufacturing systems.

At the time of writing, at least 10 companies[20] offer metal-based AM systems: 3D Systems, EOS, Concept Laser, SLM Solutions, DMG Mori, Arcam, BeAM, ExOne, Fabrisonic and Sciaky. This indicates the growth potential of metal-based AM systems in the near future. The technologies used by these companies mostly revolve around laser sintering and laser melting of a layer of powder. Certain companies like BeAM and Fabrisonic use different techniques. BeAM uses laser metal deposition (LMD) and Fabrisonic uses low-temperature ultrasonic additive manufacturing (UAM) technology.

Most of the current metal-based AM systems are not yet able to produce part quality surface finishes to the standards of CNC milling or investment casting. Nonetheless, the technology is rapidly growing and beginning to achieve the same quality as sand casting.[20] Techniques that introduce a hybrid between layer deposition and CNC milling like Fabrisonic's may solve the surface finish problem, though it adds complexity to the process. A key advantage is that metal-based AM systems can also be used to make tooling parts and moulds.

In order to increase the functionality of the 3D printed metal parts, there is a trend towards multi-material processing. This has been carried out by systems that use directed deposition process such as BeAM's LMD system and DMG Mori's SLS system. From SLM system, there are also new approaches to achieve multi-material processing, such as the method of separating two different materials within a single dispensing coating system which allows the deposition of different materials along each coating direction.[21] This has expanded the potential of metal 3D printing.

There is a general trend of steady increase in the number of publications of metal 3D printing from 1999 to 2016,[22–33] with the exception of 2008 which could be caused by the 2008 global economic crisis. Sales of AM systems for metal parts are also increasing, as shown in Fig. 9.14. An estimated 808 metal AM machines were sold in 2015, the growth of 46.9% over 2014, when 550 metal AM machines were sold. Sales of metal AM systems will continue to grow rapidly and new competitors will enter the market with new machines, which is likely to fuel a wave of new sales in the industry.

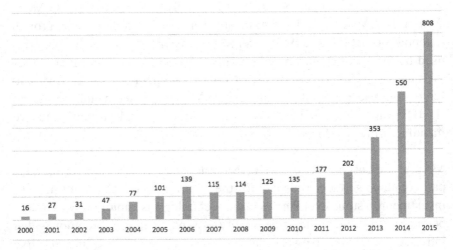

Figure 9.14: Sales of AM systems for metals parts from 2000 to 2015.
Adapted from Wohler's Report 2016.

9.3.4 *Rapid manufacturing and tooling*

RM&T has evolved from the application of AM towards direct manufacturing. It refers to a process that uses AM technology to produce finished manufactured parts or templates, which are then used in manufacturing processes like moulding or casting, to directly build a limited volume of small prototypes or components. However, it is highly unlikely that this AM-driven process will ever reach the kind of production capacity of processes such as injection moulding.[34] With the enhancement of AM technologies, an increasing number of companies from aerospace, architecture, motor sports, medical, dental and consumer products industries are turning to RM&T for customised and short-run production. The prospect of RM&T to grow and become the largest application of additive fabrication is optimistic.[34]

Typically, the costs incurred using RM&T applications (around $1000) are much lower than conventional high-end tooling cost (around $300,000). The tooling time is also notably shorter than conventional tooling time. This is mainly due to the fact that AM processes are able to fabricate models directly and quickly without the need to go through various stages of the conventional manufacturing process (i.e., post-processing). Moreover, it is more economical to build small intricate parts in low quantities using RM&T as there is compelling evidence that rapid manufacturing may be less expensive than traditional manufacturing approaches.

In addition, RM&T has the ability to make changes to the design at minimal cost even after the product has reached the production stage. Such a feature is highly desirable as it is usually very expensive to make modifications to the moulds (if the moulding process is the fabrication for the finished product) and the time needed to modify the mould is also likely to be lengthy. Thus RM&T is much preferred over the conventional tooling method in the early stages of the manufacturing cycle as modifications can be done almost instantaneously by simply editing the STL files, and the cost incurred is relatively low.

However, the downside to AM processes is its lack of an appropriate range of materials for this application. Not all materials available are suitable for all products but nevertheless, current materials, such as sintered nylon, epoxy resin and composite materials are often good enough for most low volume applications. Moreover, since RM&T has been successfully implemented in the aerospace industry, which imposes stringent quality demand, it is entirely possible that such applications can also be effectively implemented in other industries.

9.3.5 *Others*

Besides the four prominent development trends discussed, eight other trends have also been observed in the development of AM.

9.3.5.1 *Ease of use and post-processing*

One of the development trends that has been going on continuously since the first AM system is to make AM systems easier to use than before. All AM systems are supplied with user-friendly software that provides good visual feedback as well as useful automatic or semi-automatic functions to use with the 3D model and the machine.

Attempting to make post-processing easier to carry out or to eliminate it altogether is another notable development trend. As systems are able to produce parts with high dimensional accuracy and better surface finish, the need for elaborate post-processing is also diminished. Companies introduced various technologies to simplify, if not eliminate, post-processing. Stratasys introduced water-soluble supports even on their low-cost Dimension system line. With Soluble Support Technology (SST), users simply immerse the prototype into a water-based solution and wash away the supports.

9.3.5.2 *Improvement of building speed and product quality*

One other development trend is the improvement of building speed and the product quality of AM systems. As with the ease of use, this development has been continuously observed. Newer systems introduced are rapidly replacing older systems that are slow in building time and weak in accuracy. HP's multi-jet fusion (MJF) and Carbon3D's CLIP are two systems that have replaced traditional systems in terms of building speed and prototype quality.

HP unveiled two of its MJF machines in 2016: the HP Jet Fusion 3D 3200 and HP Jet Fusion 3D 4200. According to HP, these machines are able to produce parts at half the cost and up to 10 times faster than FDM and SLS machines of similar price range. Despite the speed and economic advantage, HP's MJF machines can produce parts with great accuracy and surface finishes. The HP 3200 series was introduced as the entry model and is suitable for rapid prototyping. Meanwhile, the HP 4200 series highlights its higher productivity and lower cost per part as compared to the 3200 series, thus making it ideal for short-run production as well as rapid prototyping. Both models come with the HP Jet Fusion 3D Processing Station with Fast Cooling (Fast Cooling is optional for 3200 series), facilitating material recycling and post-processing of the printed part. However, the processing station is sold as a separate unit from the main printer. More details can be found in Section 5.9.

Carbon3D's CLIP grows objects from a pool of resin by carefully balancing the interaction of light and oxygen. Traditional 3D printed parts are inconsistent as their mechanical properties vary depending on the direction the parts were printed due to the layer-by-layer approach. However, parts printed with CLIP are much more like injection-molded parts. CLIP produces consistent and predictable mechanical properties, creating parts that are solid on the inside.

CLIP process is paired with a secondary curing stage to unlock engineering properties. Traditional additive approaches to photo-polymerisation typically produce weak and brittle parts. Carbon3D's CLIP overcomes this by embedding a second heat-activated reactive chemistry in their materials. This results in high-resolution parts with engineering-grade mechanical properties. CLIP moves beyond the limitations of 3D printing to offer unprecedented speed, quality and choice.

9.3.5.3 *Standardisation of systems*

Several international standards organisations (ISO) are making significant progress toward the development of the AM standards.

The American Society for Testing and Material (ASTM) International Committee F42 on Additive Manufacturing Technologies was formed in 2009 and it aims to develop consensus standards that will support the adoption of AM across multiple industrial sectors. As of February 2016, ASTM has published 11 AM industry standards, three of them jointly with ISO.

ISO/TC 261 on AM was formed in 2011 and it aims to create standards and guidelines with different levels of detail, ranging from common standards that are quite general to specialised standards for specific materials, processes or applications. ISO/TC261 is organised into four working groups: WG 1: Terminology coordinated by Sweden; WG 2: Methods, Processes and Materials coordinated by Germany; WG 3: Test Methods coordinated by France and WG 4: Data and Design coordinated by the UK. As of February 2016, six AM industry standards had been published.

The following ASTM standards are active under F42.01 Test Methods:

- F2971-13 Standard Practice for Reporting Data for Test Specimens Prepared by AM
- F3122-14 Standard Guide for Evaluating Mechanical Properties of Metal Materials Made via AM Processes
- ISO/ASTM52921-13 Standard Terminology for AM-Coordinate Systems and Test Methodologies

The following ASTM standards are active under F42.04 Design:

- ISO/ASTM52915-13 Standard Specification for Additive Manufacturing File Format (AMF) Version 1.1

The following ASTM standards are active under F42.05 Materials & Processes:

- F2924-14 Standard Specification for AM Titanium-6 Aluminum-4 Vanadium with Powder Bed Fusion
- F3001-14 Standard Specification for AM Titanium-6 Aluminum-4 Vanadium ELI (Extra Low Interstitial) with Powder Bed Fusion
- F3049-14 Standard Guide for Characterizing Properties of Metal Powders Used for AM Processes
- F3055-14a Standard Specification for AM Nickel Alloy (UNS N07718) with Powder Bed Fusion
- F3056-14e1 Standard Specification for AM Nickel Alloy (UNS N06625) with Powder Bed Fusion
- F3091/F3091M-14 Standard Specification for Powder Bed Fusion of Plastic Materials

The following ASTM standards are active under F42.91 Terminology

- ISO/ASTM52900-15 Standard Terminology for AM — General Principles — Terminology

9.3.5.4 *New research areas and fields of AM*

Researchers and scientists are always keen in finding new areas that AM can step in to fill the gap. In 2014, researchers from the Federal Polytechnic School of Lausanne, Switzerland, the Imperial College, UK and the University of Waterloo, Canada started to carry out a study on the field of printed electronics and intelligent 3D printing. This research aims to produce intelligent multi-function components using AM.[35]

In July 2016, The Virtual Foundry launched Filament™, a new metal infused filament that enables metal printing on any FDM 3D printers. The infused filament contains 88.5% of metal powder and 11.5% of plastic which can be removed by polishing or post-processing. This new material is expected to bring the cost of metal printing down greatly and expand the capabilities of metal printing.

Meanwhile, to encourage the wider use of AM technologies, researchers have also looked into the use of AM in industries such as outer space, which NASA is currently doing, clothing, fashion, movies, food, weapons and much more. In the future, AM will form the backbone of the world's economy and it will be an integral part of industrial automation.

9.3.5.5 *Design for AM*

According to Seepersad *et al.*,[36] AM is redefining how we design, make and qualify products. For engineers, it liberates the designer by opening up the design space and removing many of the conventional manufacturing constraints. AM flips "design for manufacturing" on its head and provides "manufacturing for design", allowing us to produce geometries that were previously impossible. Meanwhile, multi-material AM systems allow designers to functionally grade material compositions to achieve unique product functionality and multipart assemblies can now be additively manufactured as a single part. Moreover, the ability to fabricate parts layer-by-layer enables lightweight structures, internal cooling passages and a host of other design benefits that increase product performance, reduce manufacturing lead-time and shorten the product-development time. AM is even changing the economics of production by reducing setup time, eliminating tolling costs and enabling small batch of runs of customised product offerings.

9.3.5.6 *Increase in AM applications in other areas*

AM technologies have been used in a broad spectrum of applications in the field of biomedical engineering, food, building and construction, marine and offshore. A variety of AM systems has been used in the production of scale replicas of human bones and body organs to advance customised drug delivery devices and other areas of medical sciences, including anthropology, palaeontology and medical forensics.[37] FDM, SLS, ColorJet Printing (CJP), formerly known as Three-Dimensional Printing technology (3DP) and SLA are the most common systems employed in the fabrication of tissue engineering scaffolds.

Intensive studies on AM techniques related to biomedical applications have been conducted in recent years. Studies in various disciplines are done to evaluate AM techniques, compare AM techniques with conventional techniques and improve the scaffold structures fabricated with various AM techniques, to customise implants and to explore new techniques/materials.[37–46] Although some success in studies for tissue engineering using AM technologies is observed, this technique has yet to deliver significant progress in the clinical use. The utilisation of AM technologies will grow as more low-cost systems with improved features are introduced to the market, thus encouraging the wider use of AM technologies.

3D printed edible items have gained more attention recently. Two of the world's largest 3D printing companies, Stratasys and 3D Systems, have begun to invest more on 3D food printing. Stratasys is developing a chocolate printer and filing for patents on processes and developing towards a commercial machine. 3D Systems has publicly declared their intentions to develop a consumer 3D printer that will make chocolate treats. 3D printing of food enables automatic food production and customisation. Further development in this area will require developing more stable print materials, higher speed of printing, greater safety controls, and so on.[47]

While more studies were carried out, AM has been widely used by architects and designers in construction. In March 2013, a Canal House was designed and built by architects from a Dutch firm called DAS.[48] This shows the capability of AM to produce prototypes on a large scale. Furthermore, China has also produced buildings via AM technology in

Shanghai in 2014. The buildings, made using 3D printers and developed by Suzhou Yingchuang Science and Trade Development Co., Ltd, look similar to buildings built using conventional methods. Even though each layer is around 3 cm tall, its hardness has been proven to be five times that of common construction materials.[49]

However, the use of AM in construction is still under further study and development, especially in terms of the accuracy and materials used. The fire resistance and durability of AM-based buildings need to be further analysed before mass production can even be considered.

For marine and offshore, 3D printed parts have become more widely accepted as it enables immediate repair and manufacture of spare parts on site. For example, experts in AM from Dshape, a pioneer in 3D printing reef units made out of non-toxic patented sandstone material, have been utilising AM in their marine and offshore study. AM enables them to make more intricate designs, which are similar to natural coral structures, and this has helped them to build an artificial reef area under the sea to protect the marine environment.[50] Many academic institutions and companies are also investing in research and development in this new technology. For example, marine and offshore is one of the key areas in which the Singapore Centre for 3D Printing (SC3DP) aims to strengthen and expand 3D capabilities.[51] Recent research on 3D printing for marine and offshore aims to improve the mechanical properties of 3D-printed metal parts and increase the build size.

Current technologies allow fabrication of not only complex physical prototypes but also functional electronic components and circuitries. It is now possible to fabricate simple circuits embedded with several kinds of passive components, for example, resistors, capacitors and inductors, which are the most fundamental building blocks of any electrical circuits. Future developments and integrations of printing electronics into a wider range of applications will continue to be researched upon.

9.3.5.7 *4D printing*

AM has been introduced since the late 1980s. Although a considerable amount of progress has been made in this field, there is still a lot of research work to be done in order to overcome the various challenges remained. Recently, one of the actively researched areas lies in the AM of smart materials and structures.[52] Smart materials are those materials that have the ability to change their shape or properties under the influence of external stimuli. With the introduction of smart materials, the AM-fabricated components are able to alter their shape or properties over time (the 4th dimension) as a response to the applied external stimuli. Hence, this gives rise to a new term called "4D printing" to include the structural reconfiguration over time.

9.3.5.8 *Bioprinting and 4D bioprinting*

3D bioprinting has been invented for more than a decade. A disruptive progress is still lacking for the field to significantly move forward. Recently, the invention of 4D bioprinting technology may point a way and hence the birth of 4D bioprinting. Compared to 3D bioprinting, there is a stimulation to trigger the as-printed tissue or organ preforms to change over time in a predefined path for 4D bioprinting. In this sense, the combination of 3D bioprinting and bioreactor could be a form of 4D bioprinting, provided the change in tissues/organ can be pre-defined. The future forms of 4D bioprinting would be really unpredictable. However, the differences between 3D bioprinting and 4D bioprinting will continue to widen when more and more research results are available. The early forms of 4D bioprinting may be just the tip of an iceberg; in addition to shape, size and pattern, there could be more other forms of changes in future, such as micro-structure, property or even functionality. The era of 4D bioprinting is on its way.

References

1. Wall, MB (1991). Making sense of prototyping technologies for product design. MSc thesis, MIT, USA.

2. Fahad, M and N Hopkinson (2013). Evaluation of parts produced by a novel additive manufacturing process. *Applied Mechanics and Materials*, 315, 63–67.

3. Atzeni, E, L Iuliano, P Minetola and A Salmi (2012). Proposal of an innovative benchmark for accuracy evaluation of dental crown manufacturing. *Computers in Biology and Medicine*, 42(5), 548–555.

4. Johnson, WM, M Rowell, B Deason and M Eubanks (2011). Benchmarking evaluation of an open source fused deposition modeling additive manufacturing system. In *Proceedings of the 22nd Annual International Solid Freeform Fabrication Symposium,* pp. 197–211.

5. Vandenbroucke, B and JP Kruth (2007). Selective laser melting of biocompatible metals for rapid manufacturing of medical parts. *Rapid Prototyping Journal*, 13(4), 196–203

6. Abdel Ghany, K and SF Moustafa (2006). Comparison between the products of four RPM systems for metals. *Rapid Prototyping Journal*, 12(2), 86–94.

7. Dimitrov, D, W van Wijck, K Schreve and N de Beer (2006). Investigating the achievable accuracy of three dimensional printing. *Rapid Prototyping Journal*, 12(1), 42–52.

8. Mahesh, M, YS Wong, JYH Fuh and HT Loh (2006). A six-sigma approach for benchmarking of RP&M processes. *International Journal of Advanced Manufacturing Technology*, 31(3–4), 374–387.

9. Grimm, TA. (2005). *3D Printer Dimensional Accuracy Benchmark.* Fort Mitchell: T.A. Grimm & Associates, Inc.

10. Gervasi, VR, A Schneider and J Rocholl (2005). Geometry and procedure for benchmarking SFF and hybrid fabrication process resolution. *Rapid Prototyping Journal*, 11(1), 4–8.

11. Mahesh, M, YS Wong, JYH Fuh and HT Loh (2004). Benchmarking for comparative evaluation of RP systems and processes. *Rapid Prototyping Journal*, 10(2), 123–135.

12. Dalgano, KW and RD Goodridge (2004). Compression testing of layer manufactured metal parts: The RAPTIA compression benchmark. *Rapid Prototyping Journal*, 10(4), 261–264.

13. Wohlers, T (2016). *Wohlers Report 2016: 3D Printing and Additive Manufacturing State of the Industry*. Fort Collins: Wohlers Association, Inc.

14. *The History of 3D Printing*. Retrieved from http://www.3ders.org/3d-printing/3d-printing-history.html

15. 3D Systems. *3D Systems' Personal 3D printers*. Retrieved from http://www.3dsystems.com/3d-printers/personal/overview

16. 3D Systems. *ProJet® 1000 & 1500 Personal 3D Printers*. Retrieved from http://www.3dsystems.com/sites/www.3dsystems.com/files/projet-1000-1500-us.pdf

17. *Price of Stratasys Moji 3D printer*. Retrieved from http://desktop-3d-printers.findthebest.com/q/28/12373/How-much-is-the-Stratasys-Mojo-3D-printer

18. Charles WH (2012). Apparatus for production of three-dimensional objects by stereolithography. *US Patent 4,575,330 A*.

19. Yap, CY, CK Chua, ZL Dong, ZH Liu, DQ Zhang, LE Loh and SL Sing (2015). Review of selective laser melting: materials and applications. *Applied Physics Reviews*, 2(4), 041101.

20. Wohlers, T (2003). *An Explosion of Metal-based RP Systems*. Wohlers Associates Inc. http://wohlersassociates.com/blog/2003/12/an-explosion-of-metal-based-rp-systems/

21. Liu, ZH, DQ Zhang, SL Sing, CK Chua and LE Loh (2014). Interfacial characterization of SLM parts in multi-material processing: metallurgical diffusion between 316L stainless steel and C18400 copper alloy. *Materials Characterization*, 94, 116–125.

22. Tan, X, Y Kok, WQ Toh, YJ Tan, M Descoins, D Mangelinck, SB Tor, K Fai Leong and CK Chua (2016). Revealing martensitic transformation and α/β interface

evolution in electron beam melting three-dimensional-printed Ti-6Al-4V. *Scientific reports*, 6.

23. Yap, CY, CK Chua and ZL Dong (2016). An effective analytical model of selective laser melting. *Virtual and Physical Prototyping*, 11(1), 21–26.

24. Bai, J, RD Goodridge, S Yuan, K Zhou, CK Chua and J Wei (2015). Thermal influence of CNT on the polyamide 12 nanocomposite for selective laser sintering. *Molecules*, 20(10), 19041–19050.

25. Lam, LP, DQ Zhang, ZH Liu and CK Chua (2015). Phase analysis and microstructure characterisation of AlSi10Mg parts produced by selective laser melting. *Virtual and Physical Prototyping*, 10(4), 207–215.

26. Wu, W, SB Tor, CK Chua, KF Leong and A Merchant (2015). Investigation on processing of ASTM A131 Eh36 high tensile strength steel using selective laser melting. *Virtual and Physical Prototyping*, 10(4), 187–193.

27. Tan, X, Y Kok, YJ Tan, M Descoins, D Mangelinck, SB Tor, KF Leong and CK Chua (2015). Graded microstructure and mechanical properties of additive manufactured Ti–6Al–4V via electron beam melting. *Acta Materialia*, 97, 1–16.

28. Kok, Y, X Tan, SB Tor and CK Chua (2015). Fabrication and microstructural characterisation of additive manufactured Ti-6Al-4V parts by electron beam melting: This paper reports that the microstructure and micro-hardness of an EMB part is thickness dependent. *Virtual and Physical Prototyping*, 10(1), 13–21.

29. Sun, G, J An, CK Chua, H Pang, J Zhang and P Chen (2015). Layer-by-layer printing of laminated graphene-based interdigitated microelectrodes for flexible planar micro-supercapacitors. *Electrochemistry Communications*, 51, 33–36.

30. Loh, LE, CK Chua, WY Yeong, J Song, M Mapar, SL Sing, Z-H Liu and DQ Zhang (2015). Numerical investigation and an effective modelling on the selective laser melting (SLM) process with aluminium alloy 6061. *International Journal of Heat and Mass Transfer*, 80, 288–300.

31. Zhang, DQ, ZH Liu, QZ Cai, JH Liu and CK Chua (2014). Influence of Ni content on microstructure of W–Ni alloy produced by selective laser melting. *International Journal of Refractory Metals and Hard Materials*, 45, 15–22.

32. Loh, LE., ZH Liu, DQ Zhang, M Mapar, SL Sing, CK Chua and WY Yeong (2014). Selective laser melting of aluminium alloy using a uniform beam profile: The paper analyzes the results of laser scanning in selective laser melting using a uniform laser beam. *Virtual and Physical Prototyping*, 9(1), 11–16.

33. An, J, CK Chua, T Yu, H Li and LP Tan (2013). Advanced nanobiomaterial strategies for the development of organized tissue engineering constructs. *Nanomedicine*, 8(4), 591–602.

34. Wuensche, R (2000). *Using Direct Croning to Decrease the Cost of Small Batch Production. Engine Technology International 2000*, Annual Showcase Review for ACTech GmbH.

35. *MGI's Ceradrop Announces Three New Orders for Printed Electronics and Intelligent 3D Printing Applications*. Retrieved from https://3dprintingstocks.com/wp-content/uploads/2014/04/MGI-PR-Three-orders-for-MGI-CERADROP.pdf

36. Rosen, DW, CC Seepersad, TW Simpson and CB Williams (2015). Special issue: design for additive manufacturing: a paradigm shift in design, fabrication, and qualification. *Journal of Mechanical Design*, 137(11), 110301.

37. Leong, KF, CM Cheah and CK Chua (2003). Solid Freeform fabrication of three-dimensional scaffolds for engineering replacement tissues and organs. *Biomaterials*, 24, 2363–2378.

38. Yeong, WY, CK Chua, KF Leong, M Chandrasekaran and MW Lee (2007). Comparison of drying methods in the fabrication of collagen scaffold via indirect rapid prototyping. *Journal of Biomedical Material Research Part B: Applied Biomaterials*, 82B(1), 260–266.

39. Yan, YN, R Wu, R Zhang, Z Xiong and F Lin (2003). Biomaterial forming research using RP technology. *Rapid Prototyping Journal*, 9(3), 142–149.

40. Wagner, M, N Kiapur, M Wiedmann-Al-Ahmad, U Hübner, A Al-Ahmad, R Schön, R Schmelzeisen, R Mülhaupt and NC Gellrich (2007). Comparative in vitro study of the cell proliferation of ovine and human osteoblast-like cells on conventionally and rapid prototyping produced scaffolds tailored for application as potential bone

replacement material. *Journal of Biomedical Materials Research Part A*, 83A(4), 1154–1164.

41. Jiankang, H, L Dichen, L Bingheng, W Zhen and Z Tao (2006). Custom fabrication of a composite hemi-knee joint based on rapid prototyping. *Rapid Prototyping Journal*, 12(4), 198–205.

42. Singare, S., L Dichen, L Bingheng, G Zhenyu and L Yaxiong (2005). Customized design and manufacturing of chin implant based on rapid prototyping. *Rapid Prototyping Journal*, 11(2), 113–118.

43. Li, X, DC Li, B Lu, Y Tang, L Wang and Z Wang (2005). Design and fabrication of CAP scaffolds by indirect solid free form fabrication. *Rapid Prototyping Journal*, 11(5), 312–318.

44. Tan, LL (2007). *FEATURE: Plugging Bone the Painless Way*. Innovation magazine. http://www.innovationmagazine.com/ innovation/ volumes/v4n3/features1.shtml

45. Hutmacher, DW, JT Schantz, CXF Lam, KC Tan and TC Lim (2007). State of the art and future directions of scaffold-based bone engineering from a biomaterials perspective. *Journal of Tissue Engineering and Regenerative Medicine*, 1(4), 245–260.

46. Ringeisen, BR, CM Othon, JA Barron, D Young and BJ Spargo (2006). Jet-based methods to print living cells. *Journal of Biotechnology*, 1, 930–948.

47. Lipton, JI, M Cutler, F Nigl, D Cohen and H Lipson (2015). Additive manufacturing for the food industry. *Trends in Food Science & Technology*, 43(1), 114–123.

48. *The World's "First" 3D-Printed House Begins Construction*. (2014). Retrieved from http://3dprintingindustry.com/2014/01/22/ worlds-first-3d-printed-house-begins-construction/

49. *China's first buildings made with 3D printing installed Shanghai*. (2014). Retrieved from http://khon2.com/2014/04/18/ chinas-first-buildings-made-with-3d-printing-installed-shanghai/

50. *Underwater City: 3D Printed Reef Restores Bahrain's Marine Life.* Retrieved from http://blogs.ptc.com/2013/08/01/underwater-city-3d-printed-reef-restores-bahrains-marine-life/#sthash.3AMs8zgm.dpuf

51. Café Aeronautique — The Inside of 3D Printing. (27 June 2015). Retrieved August 30, 2016 from http://siae.org.sg/cafe-aeronautique-the-inside-of-3d-printing/

52. Khoo, ZX, JEM Teoh, Y Liu, CK Chua, S Yang, J An, KF Leong and WY Yeong (2015). 3D printing of smart materials: A review on recent progresses in 4D printing. *Virtual and Physical Prototyping*, 10(3), 103–122.

53. Wohlers, T (2001). *Wohlers Report 2001: Rapid Prototyping & Tooling State of the Industry.* Fort Collins: Wohlers Association, Inc.

Problems

1.   What are the considerations when choosing a service bureau?

2.   What are the components that make up the total cost of a part built by a service bureau?

3.   Is there a correlation between the type of AM system and the industry in which the prototypes are used? Why?

4.   Name the reasons for benchmarking.

5.   Describe in detail the AM benchmarking methodology.

6.   How have the primary and secondary markets for AM performed over the years?

7.   In terms of system sales, would concept modellers outstrip industrial grade AM systems? Why?

8.   What are the likely trends in AM technology? Describe five such trends.

9.   Describe some significant AM milestones in the last 25 years. In your opinion, why are they significant?

LIST OF AM COMPANIES

1. Liquid-Based 3D Printer Manufacturer

3DCeram
URL: www.3dceram.com

3D Systems
URL: www.3dsystems.com

Asiga
URL: www.asiga.com

Carbon
URL: www.carbon3d.com

Carima
URL: www.carima.com

CMET
URL: www.cmet.co.jp

D-MEC
URL: www.d-mec.co.jp/eng/

DWS Systems
URL: www.dwssystems.com

EnvisionTEC
URL: www.envisiontec.com

Kevvox
URL: www.kevvox.com

Lithoz
URL: www.lithoz.com

PrismLab
URL: www.prismlab.com

Prodways
URL: www.prodways.com

Rapid Shape
URL: www.rapidshape.de

Shaanxi Hengtong
URL: www.china-rpm.com

Shanghi Union Technology
URL: www.union-tek.com

Shining 3D
URL: www.shining3d.com

Structo
URL: www.structo3d.com

Stratasys
URL: www.stratasys.com

Wuhan Binhu
URL: www.binhurp.com

Wuhan Huake 3D
URL: www.huake3d.com

ZRapid Tech
URL: www.zero-tek.com

2. Solid-Based 3D Printer Manufacturer

3DP Platform
URL: www.3dplatform.com

Beijing Tiertime Technology
URL: www.tiertime.com

BigRep
URL: www.bigrep.com

Cincinnati Incorporated
URL: www.e-ci.com

Cosine
URL: www.cosineadditive.com

EnvisionTEC
URL: www.envisiontec.com

Fabrisonic
URL: www.fabrisonic.com

INDMATEC
URL: www.indmatec.com

MarkForged
URL: www.markforged.com

MCor Technologies
URL: www.mcortechnologies.com

Stratasys
URL: www.stratasys.com

Titan Robotics
URL: www.titan3drobotics.com

Xery
URL: www.xery3d.com

3. Powder-Based 3D Printer Manufacturer

3D Systems
URL: www.3dsystems.com

3Geometry
URL: www.3geometry.com

Additive Industries
URL: www.additiveindustries.com

Arcam
URL: www.arcam.com

Aspect
URL: www.aspect.jpn.com

BeAM
URL: www.beam-machines.fr

Beijing Longyuan
URL: www.lyafs.com.cn

Concept Laser
URL: www.conceptlaserinc.com

EOS
URL: www.eos.info/en

ExOne
URL: www.exone.com

Guangdong Syndaya 3D Technology
URL: www.syndaya.com

Hunan Farsoon
URL: www.farsoon.net

InssTek
URL: www.insstek.de

Optomec
URL: www.optomec.com

Prodways
URL: www.prodways.com

Realizer
URL: www.realizer.com

Renishaw
URL: www.renishaw.com

Sciaky
URL: www.sciaky.com

Shaanxi Hengtong
URL: www.china-rpm.com

Sisma
URL: www.sisma.com

SLM Solutions
URL: slm-solutions.com

Trump Precision Machinery
URL: www.trumpsystem.com

voxeljet
URL: www.voxeljet.de

Wuhan Huake 3D
URL: www.huake3d.com

Xery
URL: www.xery3d.com

Xi'an Bright Laser Technologies
URL: www.xa-blt.com

ZRapid Tech
URL: www.zero-tek.com

4. 3D Bioprinter Manufacturer

BioBots
URL: www.biobots.io

Bio3D Technologies
URL: www.bio3d.tech

Nanoscribe
URL: www.nanoscribe.de

Organovo
URL: www.organovo.com

Ourobotics
URL: www.weare3dbioprintinghumans.org

Regenovo
URL: www.regenovo.com

RengenHU
URL: www.regenhu.com

3D Bioprinting Solutions
URL: www.bioprinting.ru

3Dynamic Systems
URL: www.bioprintingsystems.com

5. Ceramic Material Supplier
DWS Systems
URL: www.dwssystems.com

EnvisionTEC
URL: www.envisiontec.com

ExOne
URL: www.exone.com

Lithoz
URL: www.lithoz.com

6. Composite Material Supplier
3D Systems
URL: www.3dsystems.com

3DXTech
URL: www.3dxtech.com

Arevo Labs
URL: www.arevolabs.com

CRP Technology
URL: www.crptechnology.com

EnvisionTEC
URL: www.envisiontec.com

EOS
URL: www.eos.info/en

ExOne
URL: www.exone.com

Hunan Farsoon
URL: www.farsoon.com

LPW
URL: www.lpwtechnology.com

MarkForged
URL: www.markforged.com

Optomec
URL: www.optomec.com

Oxford Performance Materials
URL: www.oxfordpm.com

Prodways
URL: www.prodways.com

Sintergy
URL: www.sintergy.net

7. Polymer Material Supplier

3D Systems
URL: www.3dsystems.com

3DXTech
URL: www.3dxtech.com

Allied Photopolymers
URL: www.alliedphotopolymers.com

Arevo Labs
URL: www.arevolabs.com

Argyle/Bolson Materials
URL: www.argylematerials.com

Arkema
URL: www.arkema.com

Asiga
URL: www.asiga.com

Carbon
URL: www.carbon3d.com

CMET
URL: www.cmet.co.jp

CRP Technology
URL: www.crptechnology.com

D-MEC
URL: www.d-mec.co.jp

DWS Systems
URL: www.dwssystems.com

EnvisionTEC
URL: www.envisiontec.com

EOS
URL: www.eos.info/en

Evonik
URL: corporate.evonik.com

Kevvox
URL: www.kevvox.com

MarkForged
URL: www.markforged.com

NextDent
URL: www.nextdent.com

Oxford Performance Materials
URL: www.oxfordpm.com

Prodways
URL: www.prodways.com

Sintergy
URL: www.sintergy.net

Sintratec
URL: www.sintratec.com

Solvay
URL: www.solvay.com

Stratasys
URL: www.stratasys.com

Structo
URL: www.structo3d.com

Taulman 3D
URL: www.taulman3d.com

Victrex
URL: www.victrex.com

Voxeljet
URL: www.voxeljet.de

ZRapid Tech
URL: www.raptek.com

8. Metal Material Supplier

3D Systems
URL: www.3dsystems.com

Additive Metal Alloys
URL: www.additivemetalalloy.com

AP&C
URL: www.advancedpowders.com

Arcam
URL: www.arcam.com

Concept Laser
URL: www.concept-laser.de

Cookson Gold
URL: www.cooksongold-emanufacturing.com

EOS
URL: www.eos.info/en

ExOne
URL: www.exone.com

Fabrisonic
URL: www.fabrisonic.com

H.C.Starck
URL: www.hcstarck.com

Hoeganaes Corporation
URL: www.gkn.com/hoeganaes

LPW
URL: www.lpwtechnology.com

Material Technology Innovations (Mti)
URL: www.mt-innov.com

NanoSteel
URL: www.nanosteelco.com

Oerlikon Metco
URL: www.oerlikon.com/metco

Optomec
URL: www.optomec.com

OSAKA Titanium Technologies
URL: www.osaka-ti.co.jp

Praxair
URL: www.praxairsurfacetechnologies.com

Renishaw
www.renishaw.com

Sandvik
URL: www.sandvik.com

SLM Solutions
URL: www.slm-solutions.com

Valimet
URL: www.valimet.com

Xi'an Bright Laser Technologies
URL: http://www.xa-blt.com

9. Wax Material Supplier
3D Systems
URL: www.3dsystems.com

Asiga
URL: www.asiga.com

DWS Systems
URL: www.dwssystems.com

EnvisionTEC
URL: www.envisiontec.com

Solidscape
URL: www.solid-scape.com

Stratasys
URL: www.stratasys.com

10. Sand Material Supplier

EOS
URL: www.eos.info/en

ExOne
URL: www.exone.com

voxeljet
URL: www.voxeljet.de

COMPANION MEDIA PACK
ATTACHMENT

Multimedia is a very effective tool in enhancing the learning experience. At the very least, it enables self-paced, self-controlled interactive learning using the media of visual graphics, sound, animation and text. To better introduce and illustrate the subject of *Additive Manufacturing* (AM), an executable multimedia programme was created for this book. It serves as an important supplement learning tool to understand better the principles and processes of AM. The programme can be retrieved from this book's Companion Media Pack. To download the Companion Media Pack, please use the following access token activation URL: http://www.worldscientific.com/r/10200-SUPP. You will be prompted to login/register an account. Upon successful login, you will be redirected to the book's page; click on the "Supplementary" tab to locate the Companion Media Pack (see Fig. 1).

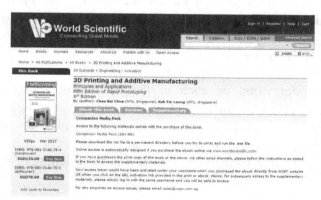

Figure 1: Screenshot of World Scientific's website to download the companion media pack. Site address: http://www.worldscientific.com/r/10200-SUPP.

More than 20 different commercial AM systems are described in Chaps. 3–5. However, only the six most matured techniques and videos on their processes are demonstrated in the multimedia pack. These six techniques and the length of their videos are listed in Table 1.

Table 1: The six AM techniques and their corresponding video lengths.

Technique	Movie Length/Min
Stereolithography Apparatus (SLA)	1:52
Polyjet	2:59
Selective Deposition Lamination (SDL)	1:14
Fused Deposition Modelling (FDM)	4:47
Selective Laser Sintering (SLS)	2:01
Selective Laser Melting (SLM)	5:06

The working mechanisms of these six methods are interestingly different from one another. In addition, the multimedia pack also includes a basic introduction, the AM process chain, AM data formats, applications and benchmarking. While the book describes the principles and illustrates with diagrams the working principle, the companion multimedia pack goes a step further and shows the working mechanism in motion. Animation techniques through the use of Macromedia Flash enhance understanding through graphical illustration. The integration of various media (e.g., graphics, sound, animation and text) is done using Macromedia's Director.

Additional information on each of the techniques, such as product information, application areas and key advantages, makes the multimedia pack a complete computer-aided self-learning software about AM systems. Together with the book, the multimedia will also provide a directly useful aid to lecturers and trainers in the teaching of the subject on AM.

COMPANION MEDIA PACK USER GUIDE

This user guide will provide the information needed to use the program smoothly.

System Requirements

The system requirements define the minimum configurations that your system needs in order to run the companion media pack smoothly. In some cases, not meeting this minimum requirement may also allow the program to run. However, it is recommended that the minimum requirements be met in order to fully benefit from this multimedia course package.

The recommended system requirements are:

- Intel Pentium PC 1.5 GHz or 100% compatible with 512 MB of RAM
- Intel Integrated 3D Graphics card
- Windows XP or higher
- 4 GB of hard disk free space (if program is installed into the hard disk)
- Sound card and speakers

The preferred system requirements are:

- Intel Pentium PC 1.8 GHz or 100% compatible with 1 GB of RAM

Installing AM Companion Media Pack

The program can be retrieved from this book's Companion Media Pack, downloadable from World Scientific's website under the Supplementary tab from http://www.worldscientific.com/worldscibooks/10.1142/9008.

Open the program by first unzipping the downloaded file, and then double-clicking the executable file start.exe to activate the courseware.

Getting Around the Program

The AM multimedia pack is a courseware developed for students, lecturers and anyone who is interested in learning more about AM. The user interface has been designed to be very simple, with a short animation introduction in order for first-time users to understand how to move around the chapters and sub-topics in the program.

The next section guides you through the various screens in the courseware and explains the functions of the icons on the screens in detail.

Main Introduction Screen

After the program has been started, a short introduction movie will be played. To skip this introduction movie, simply click on the "Skip" button. After playing the movie, you will be presented with the Main Introduction screen (see Fig. 1). When you click on the "Main Area" button shown in Fig. 1, it will direct you to the Main Menu of the courseware (see Fig. 2) where you can learn more of the key contents of the AM techniques, processes and the applications by clicking on any of the eight individual chapters desired.

Figure 1: Main introduction page.

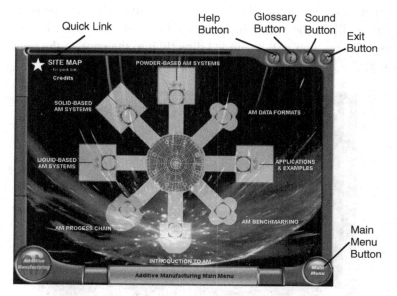

Figure 2: Main menu (after you have clicked on the "Main Area" on Fig. 1).

Choosing a Chapter

To learn more about a particular chapter, move the cursor to the word over the chapter in the homepage or in the Main Menu. The words will light up to indicate that the particular chapter can be selected. Click once on the left key on the mouse when the word is lighted up to go into that chapter.

The eight chapters are: Introduction to AM, AM Process Chain, Liquid-based AM Systems, Solid-based AM Systems, Powder-based AM Systems, AM Data Formats, Applications and Examples and Benchmarking and The Future of AM.

If you want to go to a specific sub-topic of any of the chapters, you can click on the quick link identified as "SITE MAP", which is located on the top left hand corner of the page. You will see the screen shown in Fig. 3 once you click on the words. By moving the cursor over the title of each chapter, the sub-topics of the individual chapter available for selection will be shown on the right hand side of the screen. If you click on any of them, you will be guided directly into the page of the requested sub-topic.

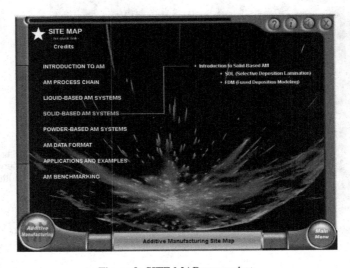

Figure 3: SITE MAP screenshot.

Graphical User Interface Layout Design

The Graphical User Interface (GUI) Layout Design (the grey border in Figs. 1 and 2) will be the same throughout the whole courseware. More advanced interactive features of this program will be covered in a later section, "Exploring the Interactive Features".

Components of the GUI Layout Design

The GUI Layout Design comprises of the main text area and various icons, as shown in the previous screen shots (Figs. 1 and 2). The main text area is where the text information of a particular chapter and sub-topic will be displayed. It is also where the user will do most of the learning.

The icons are explained as follows:

(1) *Help button.* Click on the help button to assist you in navigating the specific chapter.
(2) *Glossary button.* Click on the glossary button icon, and definitions of some terms used in the content of the chapter will be provided for your understanding.
(3) *Sound button.* This is a toggle button. Click on the sound button icon to toggle the music on or off. If the music is playing, clicking on the icon will make the music stop playing. Click on the icon again and the music will start playing again. Note that the Sound button is not active when playing any of the movies.
(4) *Exit button.* Click on the exit button to end this program and return to Windows XP. When the exit button is clicked, a confirmation screen will appear. To confirm and exit the program, click on the "Yes" button. To cancel the exit program request, click on the "No" button.

Exploring the Interactive Features

The AM multimedia courseware is an interactive, fun and lively way to learn about AM and its workings. There are many interactive features that can be found in the program. Interactive features such as animations, graphics, tables and movies are scattered around the courseware to help enhance the learning of those topics and concepts. Do enjoy and have fun exploring!

Each chapter is designed with its own theme in order to provide you with a fresh look after you finish a chapter. The section of the chapter will highlight some of the interactive features (see Fig. 4) that can be found within the courseware and how to activate them.

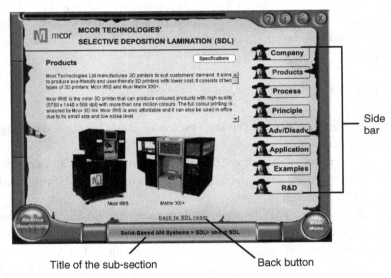

Title of the sub-section   Back button

Figure 4: Screenshot of the chapter on SDL.

Mouse Over a Hot Spot

There are many "hot spots" within the main text area in the learning screen. By moving the cursor over such "hot spots", hidden text, graphics or animations will appear (or move) on the screen.

For example, when the mouse cursor moves over "Data Preparation", the text explaining data preparation appears. There are many different "hot spots" in the entire courseware, so do explore and discover the hidden "hot spots" to help in the understanding of AM concepts in a fun and quick manner.

Buttons

In each of the six AM techniques presented (two for each of the Liquid-based, Solid-based and Powder-based AM Systems), there is a side bar for navigation within that section. Whenever the cursor is moved over any of the side bar buttons, they will be changed or highlighted in colour. Clicking once on the side bar button will bring you to the respective page. Fig. 4 shows the side bar at the "Company" page of the SDL technique.

Playing a Movie

There are six movies in this courseware to help explain the concepts of AM techniques (again, two for each of the liquid-based, solid-based and powder-based AM Systems). Whenever a particular sub-topic or page contains a related movie, the icon and words "movie clip" will appear when you move over an active icon, as seen in Fig. 5. Clicking on the movie icon will launch the movie player and play the movie automatically. You can return to the program during or after the movie is played.

The movie player has the following controls:

(1) *Play/Pause button.* Play or pause the movie at the period of time.
(2) *Forward button.* Restart the movie by clicking on the button.
(3) *Sound scroll bar.* User is able to adjust the loudness of the movie.
(4) *Movie scroll bar.* User is able to fast forward or scroll back to the part of the movie which she or he had missed out.

Figure 5: Screenshot with an active movie icon.

Self Check Quiz

There are self-quizzes in every part of the room for you to check if you have understood the content. To start the quiz, the user just has to click the "Start Quiz" button. After finishing the first round, you can choose to review the answer, retry the questions again or exit the quiz.

Figure 6: Screenshot with active quiz icon.

Printed in the United States
By Bookmasters